PUMPS   AND   PLUMBING

FOR   THE

FARMSTEAD

Prepared by
G. E. Henderson
Associate Agricultural Engineer

In Cooperation with

Jane A. Roberts
Associate Specialist in Home Electrification

Illustrator
L. H. Poole, Senior Draftsman
Drafting Service Division

Fredonia Books
Amsterdam, The Netherlands

Plumps and Plumbing for the Farmstead

by
G. E. Henderson

ISBN: 1-58963-515-9

Reprinted from the 1948 edition

Fredonia Books
Amsterdam, the Netherlands
http://www.fredoniabooks.com

PREFACE

The rapid extension of electric lines into rural territory has opened wide the field for automatic electric water systems, which in turn is developing the desire of farm people for a complete plumbing system. The need for such improvements has long been recognized as fundamental to improved sanitation and higher standards of living on farms.

In an effort to provide condensed and practical information the following text has been prepared for farm leaders and others interested in farm welfare. Since many farms do not have as much as a hand pump, and a majority of farms are still without electricity, the study has been written to cover these situations. A chapter is also included on hydraulic rams, since there are many cases where a ram would be most practical.

The text is designed for use with a teaching manual of the same name.

The writers are deeply grateful for the wholehearted cooperation of pump, hydraulic ram, and plumbing manufacturers, associations, and distributors; federal, state, and local health departments; various universities; craftsmen and technicians outside the Authority, as well as many within its group

This is the second in a series of proposed leader training courses in rural electrification. The first is "Wiring and Lighting the Farmstead."

Paralleling the above a second series of training courses is being prepared for use by leaders in extending this information to farm youth. The first of this series has been prepared and is entitled "Rural Electrification Lessons for Boys' Groups."

CONTENTS

Page

PART I

PART II

PART I

CHAPTER I

WATER SUPPLY

All of us are interested in the good health of ourselves, our families, our friends, and our neighbors. Since good water plays such an important part in contributing to our mutual well being, and contaminated water can be so destructive to health, it seems reasonable that water supply should be given first consideration in modernizing our farm homes.

In pioneer days, when the country was sparsely populated, there was little need to fear contamination even in open streams. Although there was an abundant supply of organisms in the water, they were not detrimental to health. The same is not true in rural areas today. Increased density of population per square mile has increased contamination possibilities. The use of cesspools, improperly constructed privies, in some cases a total lack of privies, and inadequately protected wells, along with certain other conditions, have all contributed to water pollution. Many rural people fail to recognize this change and are reluctant to accept the recommended sanitation practices of the state health departments.

It is not the intention in this discussion to cover the subject thoroughly, but only to point out how important it is, as a first step in modernizing the farmstead, to make certain the water supply is pure, dependable, and adequate. If it does meet these requirements it should be well protected at the surface in order to assure continued purity.

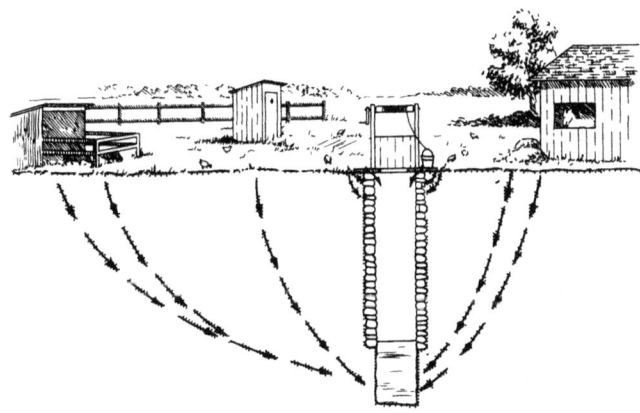

Figure 1. Common Sources of Well Contamination

Adequate Water Supply

When water is carried by hand the usage seldom exceeds 4 to 6 gallons per person per day. It is estimated that water usage increases to around 35 gallons per person per day with a completely modernized home, besides the amount consumed by livestock and for other farm uses.

From this it can be seen that many farm wells and springs have an inadequate supply to begin with, rendering useless an investment of money in an automatic pumping system, unless it happens that there is some other safe source that can be used as a reserve. In many cases a new well or cistern is necessary.

It is advisable to contact the local well driller, the health department or any other agency that is qualified to advise and possibly supply data on the availability of water in that area. Sometimes, at a relatively small cost, and old well can be made to provide an abundance of water, or a spring can be cleansed so it will develop a better flow. For great er accuracy the quantity of water flowing from a spring can be observed or measured with a weir.*

Dependable Water Supply

Dependability is a difficult feature to determine except through use or a constant period of pumping. Water is frequently supplied from ground water near the surface, which means the water level will quite likely fluctuate with rainfall. In some cases if a source is tapped at a lower level, as a result of penetrating through an impervious rock or soil layer, there is a greater possibility that the reservoir of water will be larger, more constant, and little, if at all, affected by rainfall.

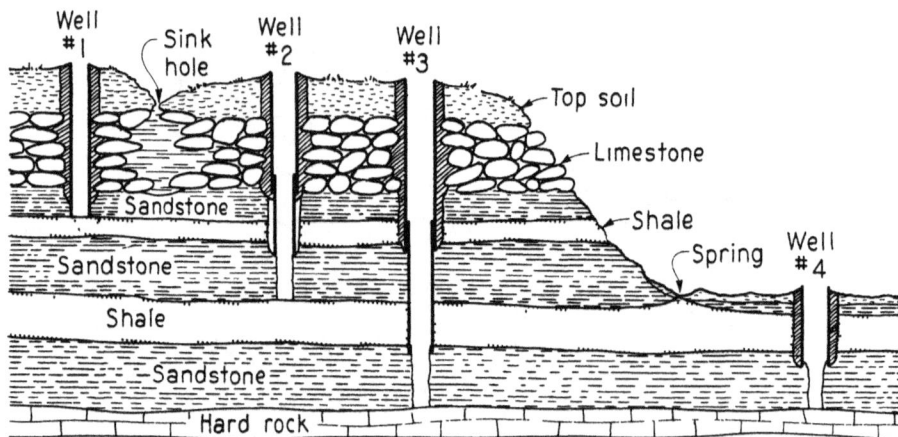

Figure 2. Well No. 1 may be dangerous. The limestone layers do little filtering and surface water penetrates easily. Wells No. 2 and 3 are probably safe. The shale is effective in keeping the first, second, and third water layers separated. However, if wells 2 and 3 are not properly grouted, water could pass from the first level down to the second or even the third level. Well No. 4 should be good.

*Note discussion in Appendix on "Use of a Weir."

Checking on other wells in the area may give an indication of the dependability of the water supply at different levels, but it is only an indication.

If a spring tends to increase in flow during a rainy season and decrease during a dry season, it is almost certain to be fed by a source near the surface and is unsafe.

Freedom From Pollution

The purification of surface water is accomplished by the filtering of water through soil, which gradually filters out the polluted matter. There is considerable difference in types of soils and their ability to filter; therefore, it is well to secure recommendations from the state health department as to how far a well should be from a barn, privy, cesspool, septic tank, or other source of possible contamination. Recommendations vary from 50 feet to 300 feet, depending on slope, type of soil, type of well, and the degree of danger from a contaminating source

If there is a suspicion that water from a sinkhole, septic tank, or other source of pollution is seeping through to the well, it can be roughly checked by emptying one pound or more of potassium permanganate into the source of polution and observing whether or not the well water turns a purple color within a few days. However, potassium permanganate is limited to use where the source of pollution is close to a well. Sometimes a stronger coloring agent is needed, as is often the case in limestone areas where pollution has been known to follow along rock crevices for a mile or more; in this case fluorescin is used. Two ounces is usually sufficient. In concentrated solution it is reddish in color, and in weak solution it appears light green. Tests of this type are not conclusive and should not replace bacteriological examinations made by health departments.

The precautionary measures listed thus far take care of water that falls on the ground surface and filters through to the well, but to complete the precautions the well must be protected at the ground level with a well constructed platform elevated above ground level, and with the pump on a base or casing above the platform. Provision should also be made to keep the water from running under the platform into the well. Likewise, springs should be walled and covered to protect them against surface water and the possibility of dirt and small animals getting into the spring box. An overflow spout should extend from the box to avoid dipping of utensils into it. Details and drawings are available from various state health departments and universities.

It is good practice to have the water of wells and springs checked occasionally to make certain it is free from pollution. There is the common misconception that water which is tested and shows a degree of

pollution indicates the presence of typhoid fever germs. This is not
generally true, since most tests are conducted to determine the presence
of a certain group of bacteria found in the intestinal tracts of warm
blooded animals. When bacteria of this type are not present it is probable the water is well filtered and free from dangerous organisms. However, to be absolutely certain several tests are required at three- or
four-week intervals over a period of several months. Wells that show
polluted water at some seasons of the year may test pure at other seasons.

Open dug wells are so certain to be polluted that most health departments refuse to check the water unless the wells are first covered according to approved practice.

Figure 3. Incorrect and Correct Types of Spring Protection

Cisterns

In sections where an adequate water supply is difficult to obtain, or the
supply is of questionable character, a cistern may be desirable. It may
be used as a supplementary or as a full-time water supply if of sufficient
capacity.

The cistern should be watertight to prevent loss of water and inflow of
ground water. The practice followed in some sections, of digging a hole
in the ground and plastering the walls and floor with a cement plaster

and no reinforcements, does not assure a watertight cistern. A good concrete structure is probably best.

In figuring the size of a cistern, the probable usage and the distribution of rainfall throughout the year should be taken into consideration. It is good practice to build one large enough for a six-month storage; however, in some sections a three-month storage capacity has been found sufficient. In any case, a reserve capacity should be allowed.

Figures for determining the family needs are given in chapter V of this text. Figures showing cistern dimensions to store various quantities of water are given in table XIII of the appendix. The amount of rainfall and distribution throughout the year should be secured from the local weather bureau.

Provision should be made in the downspout to divert the first rainfall so that it will not enter the cistern and pollute the accumulated water with dirt from the roof. Diverting the first rainfall is even more important if a filter is used to avoid accumulation of leaves, bird droppings, etc. If such material is allowed to enter the filter it becomes an excellent incubating medium for bacteria and the desired effect of a filter is lost.

The fact that filter material should be washed or replaced frequently to be effective, and that few people take this precaution, has caused some health departments to recommend automatic roof wash devices similar to the one shown in figure 4b. The drum or tank should be sufficiently large to catch enough water to wash the roof thoroughly. When the container is filled the remaining water flows into the cistern. By leaving the faucet open slightly the tank will drain itself between rains.

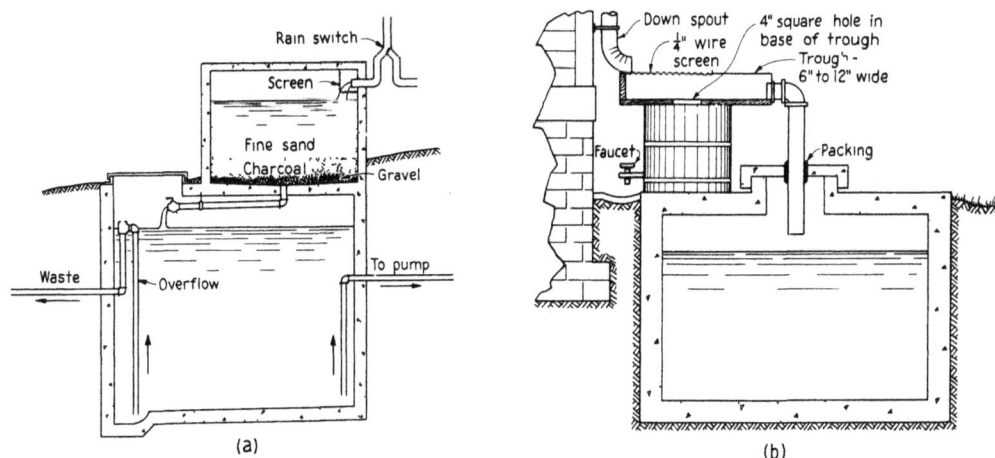

Figure 4. (a) Simple Type Charcoal Filter
(b) Automatic Roof Wash Device

CHAPTER II

ATMOSPHERIC PRESSURE--SUCTION

Water pumping reduced to its simplest terms is nothing more than a suction and pressure operation. In fact, it can be reduced still further because when water rises by suction it does so because of the force or pressure exerted by the atmosphere. Consequently, suction also becomes a pressure operation.

Atmospheric Pressure and Its Relation to Suction

Most of us realize but scarcely stop to consider that the great blanket of air surrounding us, and extending above us for several miles, exerts a tremendous pressure all over the earth's surface. At sea level the pressure exerted is 14.7 pounds per square inch. On a mountain top air pressure decreases because the blanket of air is not as thick. Our bodies are fortified against it to the extent that if the atmospheric pressure were suddenly removed our bodies would rupture because of inside pressure.

In this discussion our interest is in how atmospheric pressure can be put to work forcing water out of a well and up to the ground surface. We commonly speak of pumping water by suction, but let us see what actually happens.

Assume we have a long glass pipe extending into a dug well (figure 5). If the pipe is open at both ends, the water level in the pipe will be the same as in the well because the atmosphere presses down just as hard on the water inside the pipe as it does on the water outside in the well proper. The result is no difference in water levels.

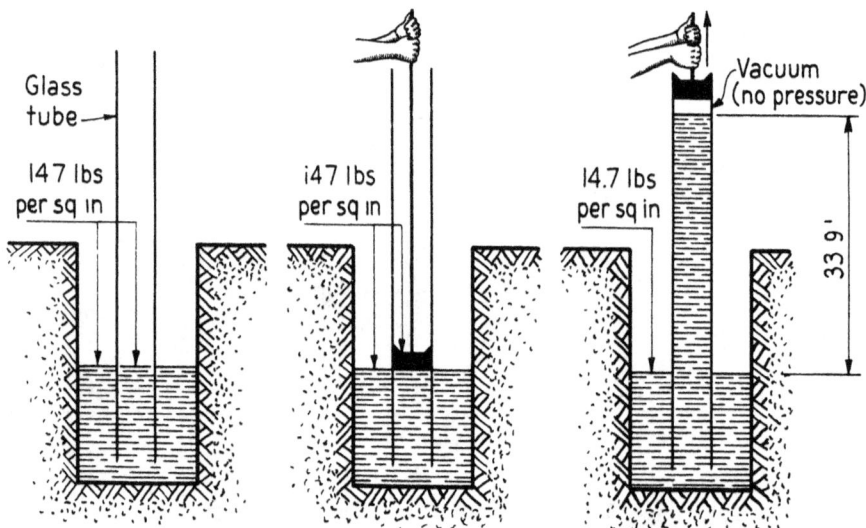

Figure 5   Figure 6   Figure 7

It would seem reasonable to assume that if the atmospheric pressure could be reduced inside the glass pipe, the air pressure on the water outside would cause the water to rise in the pipe. To do this let us fit an airtight piston into the glass pipe (figure 6) and exhaust the air by starting the piston at the water level and pulling it up. You will note from figure 7 how the water has followed the piston for 33.9 feet and then stopped. The space between that level and the top position of the piston is not an air space but a vacuum. The reason the water did not continue to follow the piston is that the water column grew heavier as it rose, until it pressed back on the water in the well with as much pressure as the atmosphere was exerting to push it up. That means that atmospheric pressure has capacity to lift water 33.9 feet, or in other words, it requires a water column 33.9 feet high to balance a pressure of 14.7 pounds per square inch.

A vacuum does not have capacity to "suck" as is the common belief, for, if it had, the water column would have continued to follow the piston past 33.9 feet. All a suction does in this illustration is to create a vacuum so that the atmospheric pressure is "put to work" in forcing the water up the tube.

It was stated that water can be raised 33.9 feet, which is theoretically true, but because of pump losses and pipe friction, which result in only a partial vacuum, practical suction lift is usually limited to 22 feet, with some manufacturers claiming as much as 28 feet.

Table I below indicates practical suction lift at different elevations.

TABLE I

Suction Lift of Pumps with Barometric Pressure
at Different Altitudes When
Pumping Cold Water

Altitude	Barometric Pressure	Equivalent Head of Water	Practical Suction Lift of:		
			Reciprocating Pumps	Centrifugal Pumps	Centrifugal Turbine Pumps
Sea level	14.70 lbs per sq in	33.95 ft	22 ft	15 ft	28 ft
1/4 mile (1320 ft) above sea level	14.02 lbs per sq in	32.38 ft	21 ft	14 ft	26 ft
1/2 mile (2640 ft) a.s.l.	13.33 lbs per sq in	30.79 ft	20 ft	13 ft	25 ft
3/4 mile (3960 ft) a.s.l.	12.66 lbs per sq in	29.24 ft	18 ft	11 ft	24 ft
1 mile (5280 ft) a.s.l.	12.02 lbs per sq in	27.76 ft	17 ft	10 ft	22 ft
1-1/4 mile (6600 ft) a.s.l.	11.42 lbs per sq in	26.38 ft	16 ft	9 ft	20 ft
1-1/2 mile (7920 ft) a.s.l.	10.88 lbs per sq in	25.13 ft	15 ft	8 ft	19 ft
2 miles (10560 ft) a.s.l.	9.88 lbs per sq in	22.82 ft	14 ft	7 ft	18 ft

Pumps limited to drawing water from depths indicated in table I fall into a definite classification called shallow well pumps.

If water is to be lifted from greater depths than those indicated, the pumps are classified as the deep well type. Both types will be studied later.

Developing Suction

Cylinder and Piston

Farm pumps operated by hand are of the cylinder and piston type, as are most of the present automatic farm water systems. The suction is created by a piston or plunger working back and forth inside a cylinder. To illustrate, note in figure 8a the downward movement of the plunger in the cylinder. As it moves down, the water below the plunger is forced through a valve in the plunger and into the upper part of the cylinder.

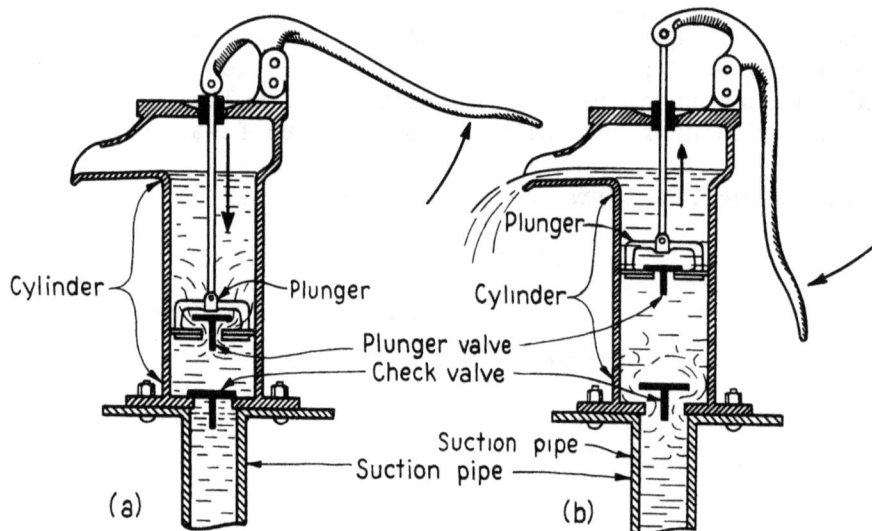

Figure 8. Common Cistern Pump

In 8b the plunger movement is up, which causes the plunger valve to close, because of the weight of the water above the valve and of the added weight of a column of water below the plunger valve. At this period of operation suction is created. In other words, when the plunger moves up it would leave a vacant space (partial vacuum) if it were not for atmospheric pressure which presses on the water in the well, causing the water in the suction pipe to force open the check valve at the bottom of the cylinder and occupy the space below the plunger.

When the plunger starts its downward movement the vacuum effect is lost
in the cylinder, and both atmospheric pressure and the weight of the
water in the cylinder press down on the water in the suction pipe. The
only thing that keeps the water in the suction pipe from returning to
the well is the downward pressure that closes the check valve at the
bottom of the cylinder. This action prevents atmospheric pressure from
being effective on the water column so that it remains in position.

If the check valve should fail to seat or if air should enter the suction
pipe through a loose pipe joint, the water column would immediately return to the well, and the pump is said to have "lost prime." In order to
pump successfully the trouble must be corrected and the water column re-
established.

Suction by Rotary Gear

The rotary gear pump is not commonly used on farms. However, there have
been a few installations and the principle of operation is near enough
to that of the piston and cylinder type that it should be understood.

Figure 9 illustrates the simple construction of the gear
pump. It is simply two gears
meshing together inside a
housing. However, it makes a
difference which way the gears
rotate. Note that they rotate
so that the teeth disengage
and open next to the suction
pipe, leaving the space between the teeth in a state of
partial vacuum. Again atmospheric pressure comes to the
rescue and forces water up the
suction pipe into the void
spaces between the teeth of
each gear. As the gear continues revolving, the water
is automatically imprisoned

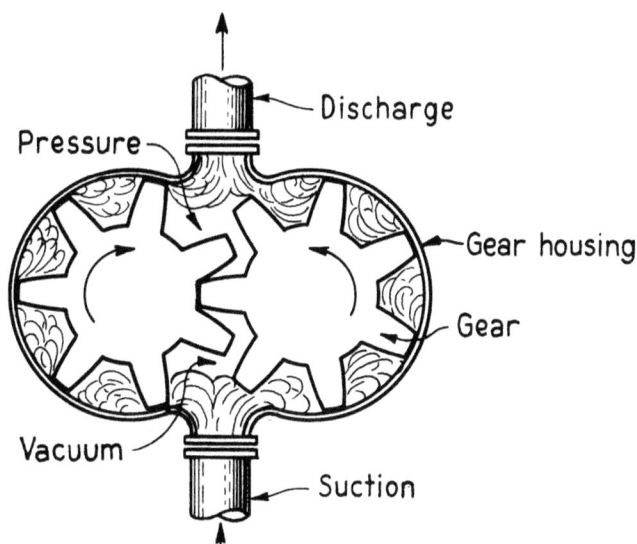

Figure 9. Gear Pump

between the gear teeth and the gear housing. When the teeth revolve to a
point where they begin to mesh, the water is squeezed out and up through
the delivery pipe.

Suction by Centrifugal Force

Centrifugal pumps are commonly used for pumping large quantities of water,
as for irrigation. However, certain variations and combinations of the
centrifugal principle are used in automatic farm water systems.

Although the rotary gear and centrifugal mechanism may appear to have much in common, they are really quite different in operating principle. Where the rotary gear has an expanding and squeezing action, the centrifugal pump is dependent on a throwing or centrifugal action.

Most of us have observed an automobile pull from a muddy road on to a hard surface road and have noted how the mud was thrown off the tires with greater and greater force as the car gained speed.

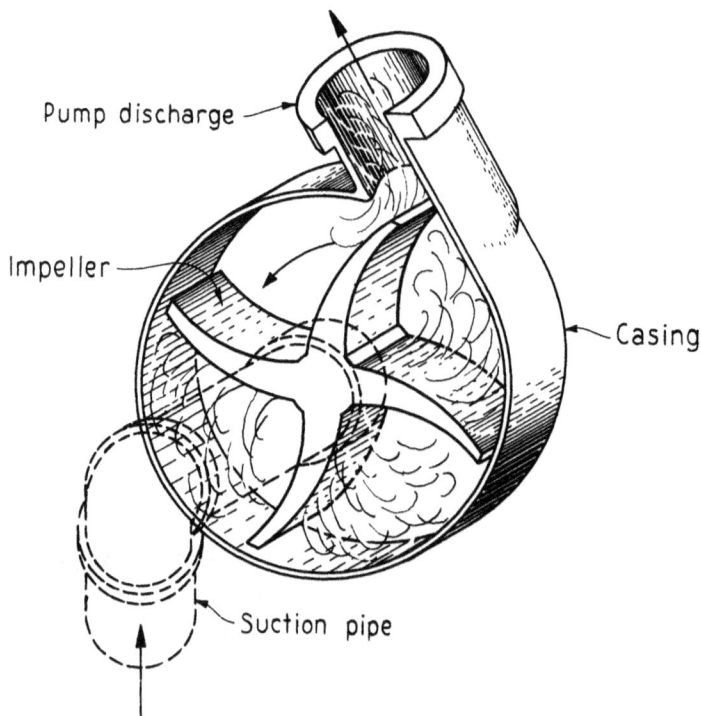

Figure 10. Centrifugal Pump

The centrifugal pump, as shown in figure 10, works on the same principle. As the wheel or impeller gains speed its throwing action, or in other words its centrifugal force, is increased. You will note that there is a housing around the impeller so that no air can enter, which means that if the water in the housing is thrown out a partial vacuum is created. Since the water is thrown to the outer edge of the impeller blades, it seems natural that the suction line should feed water to the hub of the impeller. This is actually what happens; as the impeller throws water out the discharge, atmospheric pressure pushes more water up the suction line and into the impeller. Naturally this type pump must be high speed, and it must be completely primed before it will pump.

Suction by Water Jet

Suction by water jet is not a new principle, but it is relatively new to farm water systems. As will be studied later, it is used mostly in combination with a centrifugal pump for wells up to about 80 feet deep.

Note in figure 11 that water under fairly high pressure is forced through a restriction in the form of a jet. We know from our experience with sprinkler hose that when a finger is placed over the outlet, the water

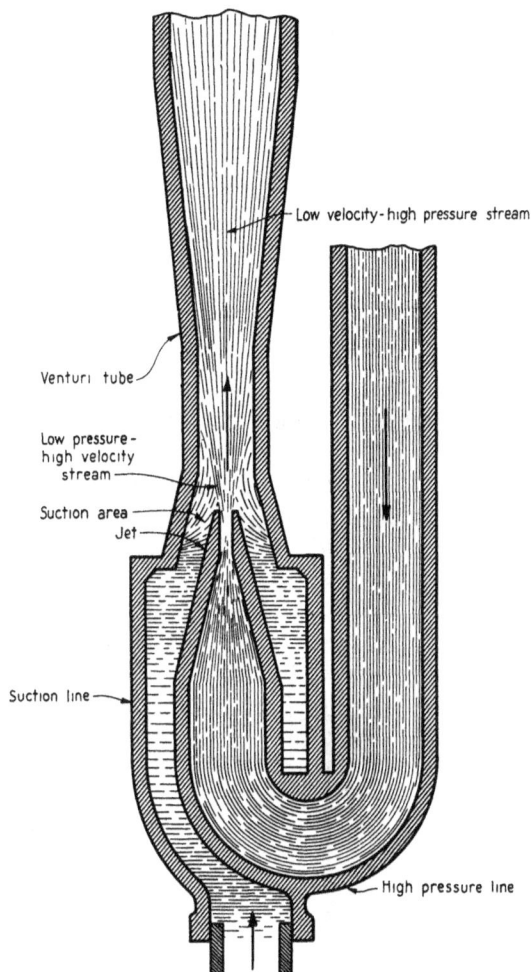

Figure 11. Ejector, or Water Jet Pump

squirts, or in other words attains velocity, which carries it several times farther through the air than when the end of the hose is left unobstructed.

The same principle holds with the jet, only in this case, it is surrounded by water instead of air. As the water under high pressure passes through the jet it attains a high velocity and the stream presents a rather irregular surface which through impact causes particles of the adjoining layer of water to be carried along and incorporated into the high velocity jet stream. The jet stream along with the additional water collected passes on up into the Venturi tube where the velocity diminishes and a high pressure again exists. The Venturi tube is a cone-shaped opening immediately above the jet where the water gradually converts from a high velocity stream at low pressure to a low velocity stream at high pressure. Of course, with water being removed in the suction chamber there is a tendency to create a vacuum and again atmospheric pressure pushes more water up to take the place of that which the jet stream removed.

CHAPTER III

PUMPS

A farm pump may be required to perform any one, two, or possibly all three of the following hydraulic tasks:

1. Lift water by suction (figure 12a).

2. Lift water by direct mechanical force by pushing water up a drop pipe to the ground level (figure 12b).

3. Force water, after it has reached ground level, to a storage tank or point of usage (figure 12c).

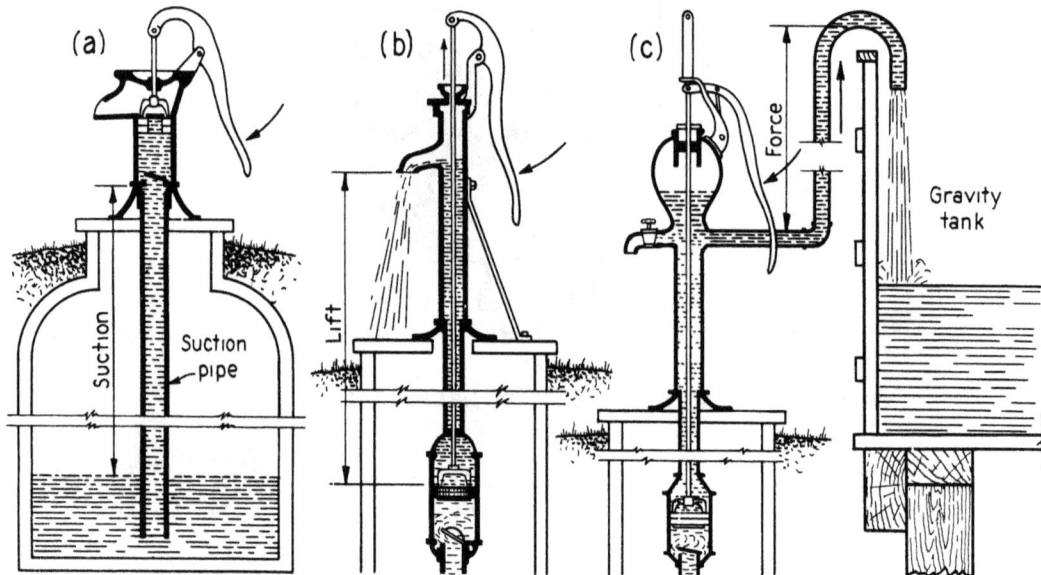

Figure 12.   Three Hydraulic Tasks a Pump May
Perform Separately or in Combination

Naturally, the design of a pump must vary in relation to the type of job to be done. A great many installations require a pump that will accomplish all three tasks. The combination of a variety of tasks, plus the various ways of doing them, as indicated in chapter II, has caused manufacturers to develop a wide selection of pumps. For sake of brevity our discussion must necessarily be confined to the more common types.

SHALLOW WELL PUMPS

Piston and Cylinder Type

Hand Operated Piston Pumps

House Suction Pump (Single Acting). The type pump illustrated in figure 12a is one of the simplest and most commonly used in farm homes.
It is designed to do one job only--that of lifting water by suction*
to the level of the pump. It is usually limited to a 22-foot vertical
lift, with a possible maximum of 25 feet. In many areas the open top
pitcher pump is in common use. It should be discouraged in favor of
the closed top, which is much more sanitary.

Force Pump (Single Acting). The shallow well force pump is the next
advance step in hand pumps. It is designed not only to lift water by
suction but to deliver it past the pump to a storage or point of usage.

The operating principle is exactly the same as that of the house suction
pump; however, the design varies in that it has a stuffing box, a cock
spout, and quite often an air chamber and a check valve in the delivery
line. The operation of these features will be discussed later.

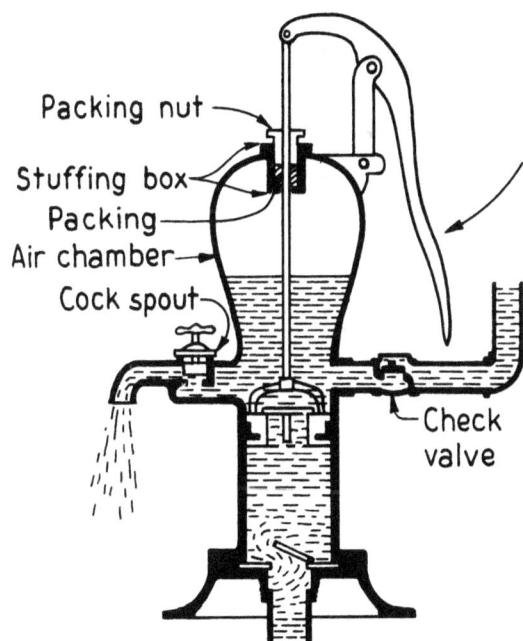

Figure 13. Single Acting, Hand
Operated Force Pump

Figure 14. Double Acting, Hand
Operated Force Pump

*As discussed in chapter II, suction is actually atmospheric pressure at
 work. In the sense that many people regard "suction" the term is probably misleading. However, its use is so general with pump men and farmers
 that it will be used during the remainder of the discussion.

Force Pump (Double Acting). In the same classification is another pump
which accomplishes the same tasks and lifts water only on the up stroke
of the plunger but discharges on both strokes. The same principle is
employed with deep well pumps (figure 24) using differential cylinders,
which will be studied later.

A more common type double acting pump is shown in figure 14. Note that
the cylinder is horizontal and that in figure 15 it has four valves instead of two. It differs from the type first described in that it not
only discharges on each stroke but also lifts water by suction on each
stroke.

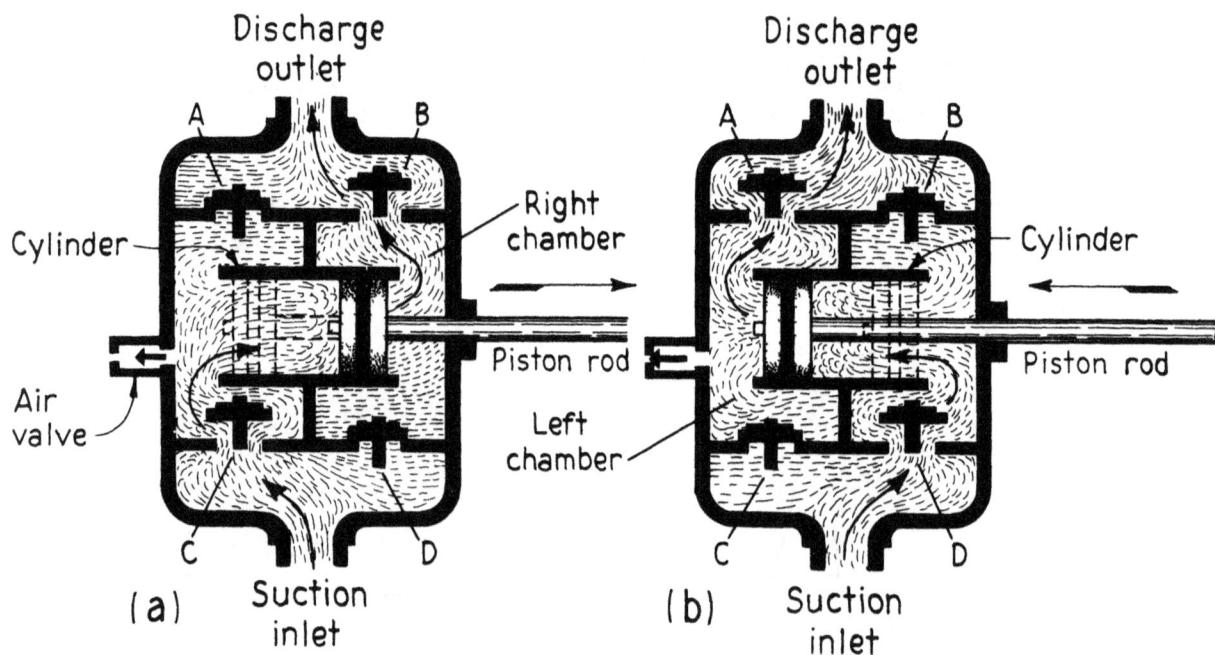

Figure 15. Operating Principle of a Double Acting Pump

To understand how this is done note that the movement of the piston in
figure 15a is from left to right, causing the water in the right chamber
to be forced out of valve B. At the same time the piston movement tends
to create a vacuum in the left chamber, causing atmospheric pressure to
force open valve C and fill the left chamber with water.

On the return stroke valves B and C close and valves A and D are forced
open (figure 15b). The double acting pump gives a more continuous flow
of water through the discharge pipe than the single acting pump, but
there is still a pulsating effect on both the inlet and the discharge.
This must be offset in certain installations. How it is accomplished
will be discussed later.

Power Operated Piston Pumps

It is possible to connect the hand operated, double acting pump just discussed with a pump jack and operate it with a gasoline engine or an electric motor. However, the general practice is to replace such a pump with one designed for power operation.

Reciprocating Pumps

The term "reciprocating" is quite commonly used to describe the standard piston type, power driven pump. The principle on which the pumping mechanism works is exactly the same as with the double acting, hand driven pump, although both outward and inward appearances vary considerably with different makes of pumps.

Since pumps of this type are in common use, we shall take time to study how they are made.

Water End. A reciprocating pump has a speed of approximately 200 to 500 revolutions per minute as compared with a hand pump at 35 to 50 strokes per minute. The higher speed greatly increases the pump capacity, but it requires an improvement in design to take full advantage of the speed.

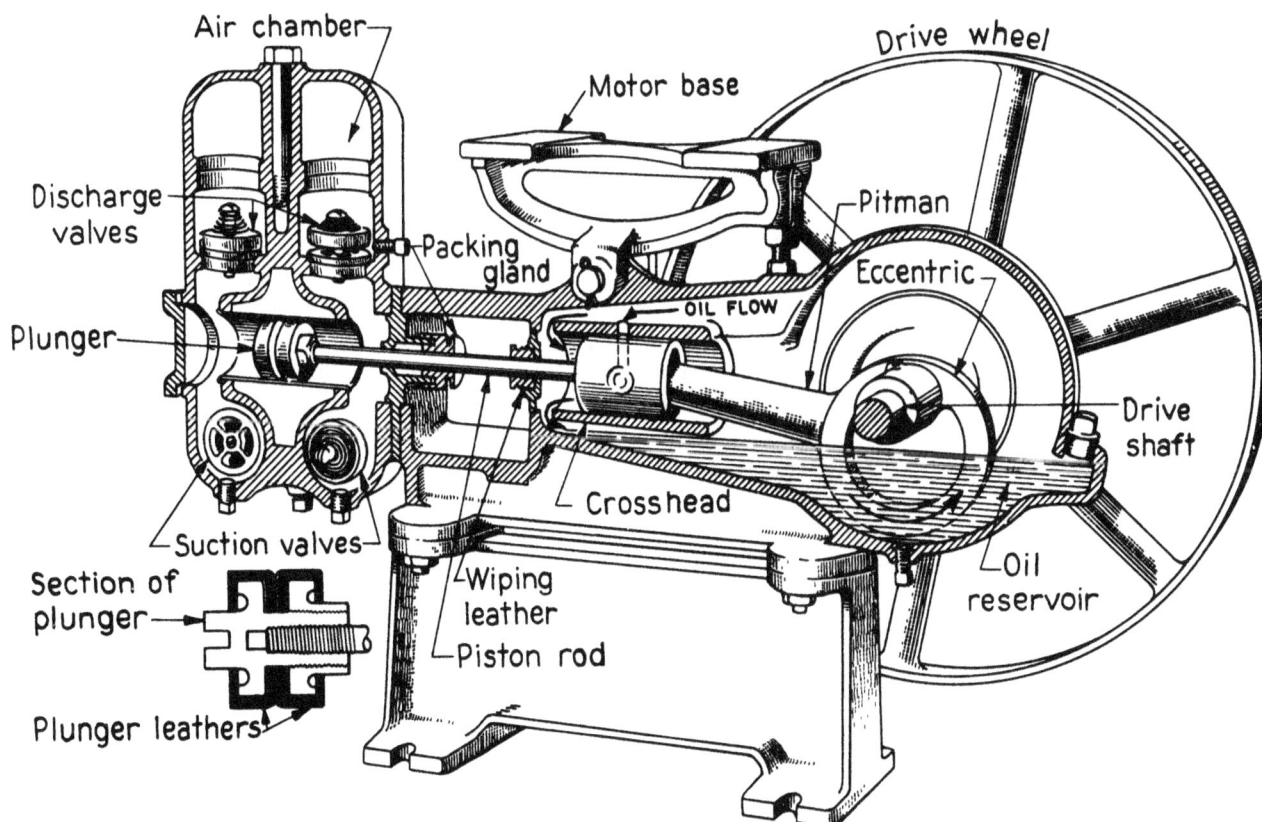

Figure 16.   Cross Section of a Shallow Well, Power Driven Piston Pump

Figure 16 is a cross section of a shallow well reciprocating pump. Note
that the pump end on the left is very similar to the double acting hand
pump just studied. There are two suction valves at the bottom and two
discharge valves on top, with a piston or plunger in the center.

The plunger is designed with two cup-shaped leathers on a metal core.
The cups are set back to back so that the rims of the leather cups are
exposed to water pressure when the pump is in action. When the plunger
stroke is to the left the rim of the left leather expands, causing it to
fit very closely against the cylinder wall and forcing the water ahead of
it without permitting it to pass the plunger. On the return stroke the
right leather expands and prevents water passage. When there is sufficient wear that the leathers permit water passage, the pump loses a portion of its pumping capacity. The leathers can be easily replaced.

Since the leathers rub against the cylinder walls it is necessary that
the cylinder be very smooth in order to avoid excessive wear. This is
accomplished by using a smooth brass liner.

The valves are equally important, for if they are slow acting or do not
seat well, water returns past them and the efficiency of the pump is
decreased. To speed up valve action a spring is used with each valve.
At the completion of a stroke the valve seats quickly and is not totally
dependent on the back pressure of water. A tight fitting valve is secured
by using a rubber valve on a bronze or brass seat.

The piston rod which imparts motion to the plunger is usually of brass,
bronze, monel metal, or stainless steel. These metals present a smooth
surface, which is necessary because of the movement of the rod through
the pump packing in the stuffing box.

The stuffing box is part of the outside of the water chamber through
which the piston rod works and which prevents the water from escaping
out of the water chamber along the piston rod. It consists of a gland
around the piston rod which is stuffed with a leather packing, graphite
packing, or asbestos impregnated with rubber. The outer end of the gland
is equipped with a nut for tightening the packing so it will fit closely
around the piston rod. Some manufacturers use a spring, which more or
less automatically keeps the packing tight.

Driving Mechanism. Most pumps of this type are belt driven, the power
being delivered to a pulley on the outside of the housing.

The drive shaft to which the pulley is attached extends through the housing, where it is supported by a bearing, and to the center of the housing,
where it supplies power to a pitman usually by means of an eccentric or
crank shaft. In some pumps the shaft extends to the other side of the
housing, where it is supported by a second bearing. The bearings are usually bronze or roller bearings. High pressure pumps use a set of speed
reduction gears ahead of the crankshaft.

From the eccentric or crankshaft, power is supplied through a steel or brass pitman rod to a crosshead, usually made of bronze or brass. The crosshead in turn supplies power to the piston rod and acts as a guide in giving the piston rod a straight back-and-forth motion.

Some provision is made on all pumps of this type to keep any leakage of water past the stuffing box from following the piston rod back to the driving mechanism. This may be in the form of a second stuffing box on the power end, a specially designed wiper that presses against the piston rod, or a deflector washer mounted on the piston rod.

Lubrication is supplied to moving parts by the action of the eccentric or crankshaft dipping into a reservoir of oil and throwing it directly on the moving parts, or throwing it to a higher elevation where it flows by gravity to moving parts. For this reason it is important that the pump pulley rotate in the direction indicated by the manufacturer.

One manufacturer has incorporated the driving mechanism into the water chamber and uses water for lubrication.

Centrifugal Type

Common Centrifugal Pump

Centrifugal pumps of the type shown in figure 10 have not been widely used on farms. This may be largely because of their limited ability to lift water by suction as indicated in table I and their ease of losing prime.

The impeller on this type pump usually differs from the "open" type shown in figure 10. Instead of the open blades revolving in the pump housing, they are designed with sides which revolve as part of the impeller, thus eliminating water slippage around the sides of the blades. This is known as a "closed" type impeller, and is illustrated in figure 17. However, the operating principle is exactly the same as discussed in chapter II.

Centrifugal Jet Pump (Shallow Well)

Some manufacturers are building shallow well centrifugal pumps with a built-in jet or ejector as shown in figure 17. The arrangement develops greater suction lift and assists in developing pressure on the discharge side.

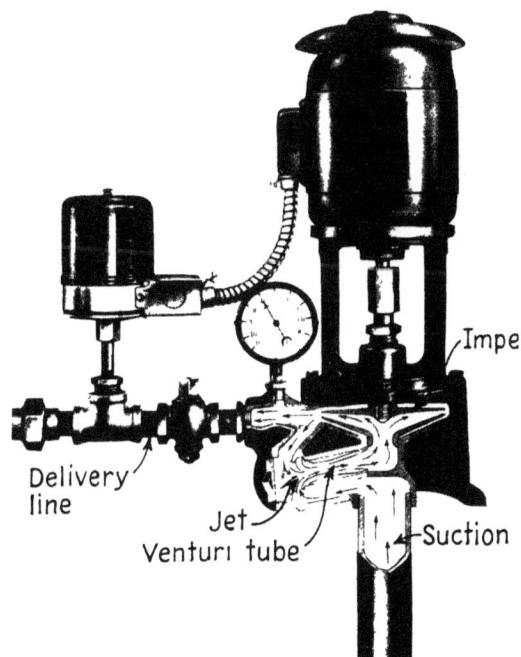

Figure 17. Centrifugal Jet Pump for Shallow Wells

Turbine Pump

Centrifugal pumps are called turbines when they are vertically mounted with one or more impellers located in the well for deep well operation. The use of this type pump is not common on farms except in irrigation districts.

The pump we shall study is the shallow well turbine which is in common use for providing general farm water supplies from shallow wells.

It is of somewhat different design from the centrifugal pump just discussed. It has only one impeller and instead of the water entering near the hub of the impeller and being thrown out the discharge by direct centrifugal action, as explained in chapter II, in this case it enters a channel near the outer edge of the housing. A series of fast moving blades move along the open side of a U-shaped channel.

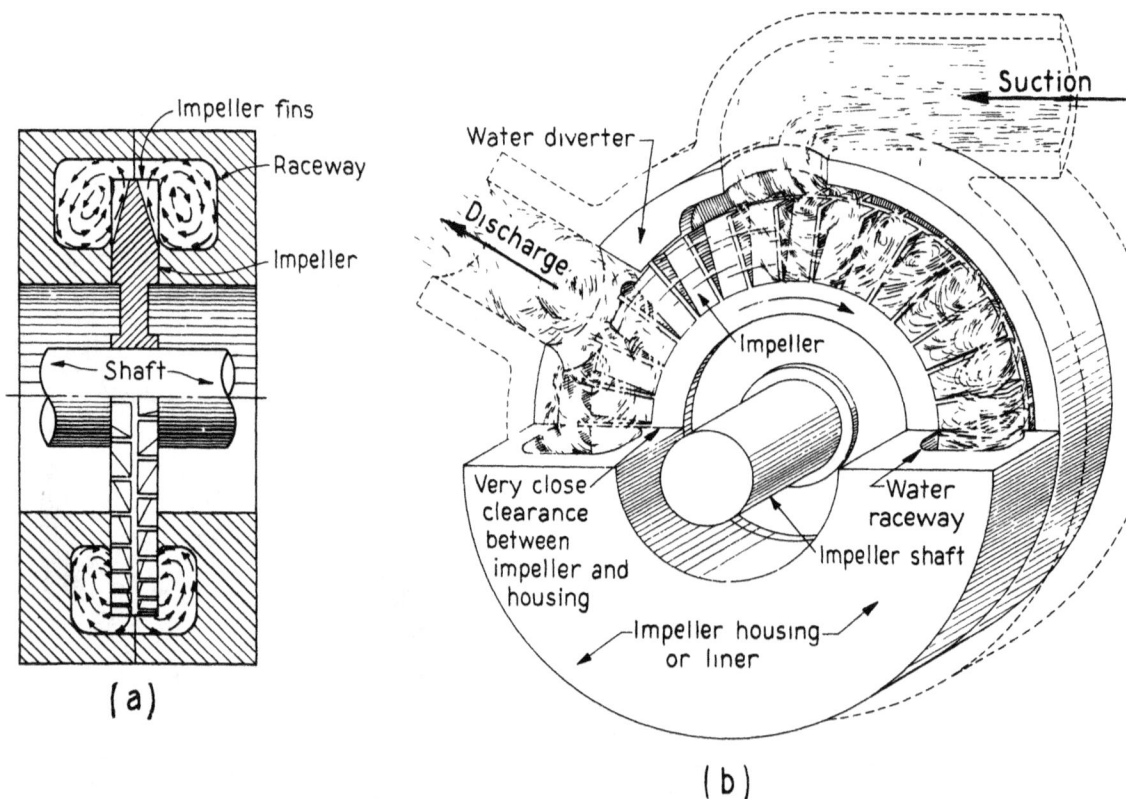

Figure 18. Pumping Action of a Single Impeller Turbine Pump
  (a) Cross Section Showing Water Movement When Thrown Off Impeller Blades
  (b) Side View Showing Relation of Suction and Discharge

The water between the blades is carried forward and thrown from the outer edge of the blades against the outer wall of the channel (figure 18a).

Since the blades throw off a continuous stream of water, that which is thrown off tends to follow the circular wall and return to the blades where it is again thrown to the outer side of the channel. Such action gives all the water in the channel a rapid, whirling, spiral movement as well as a forward movement so that it gradually increases in pressure until it is forced out the discharge (figure 18b).

A turbine pump so designed will lift water by suction from depths greater than the common centrifugal pump. In fact, manufacturers claim as much as 28 feet practical suction lift. (Note table I.) The pump will also pump air along with water, which is not true of the ordinary centrifugal pump.

Water End. The water end consists of a bronze impeller with bronze or monel metal blades. In the case of a turbine pump the impeller rotates with very close clearance opposite a bronze block containing the channels or between two bronze blocks each containing a channel. A check valve is mounted in the assembly on the suction side of the pump to hold prime.

Driving Mechanism. The impeller is directly connected with the motor by means of a stainless steel shaft. The shaft extends from the pump housing through a packing gland usually filled with a graphite or semi-metallic packing. The packing gland nut is kept loose enough to permit an occasional drop of water for lubrication. Some manufacturers use a carbon wearing ring in place of packing.

Rotary Type

In chapter II we studied the principle of suction of the rotary type pump. The two gears are the only moving parts of the pump. On the end of one of the shafts a pulley can be attached or it can be directly connected.

A stuffing box with spring compression, similar to the type used on turbine pumps, is mounted on the casing through which the drive shaft extends.

The shafts are usually stainless steel and the gears bronze.

Lubrication is with water or grease cups.

DEEP WELL PUMPS

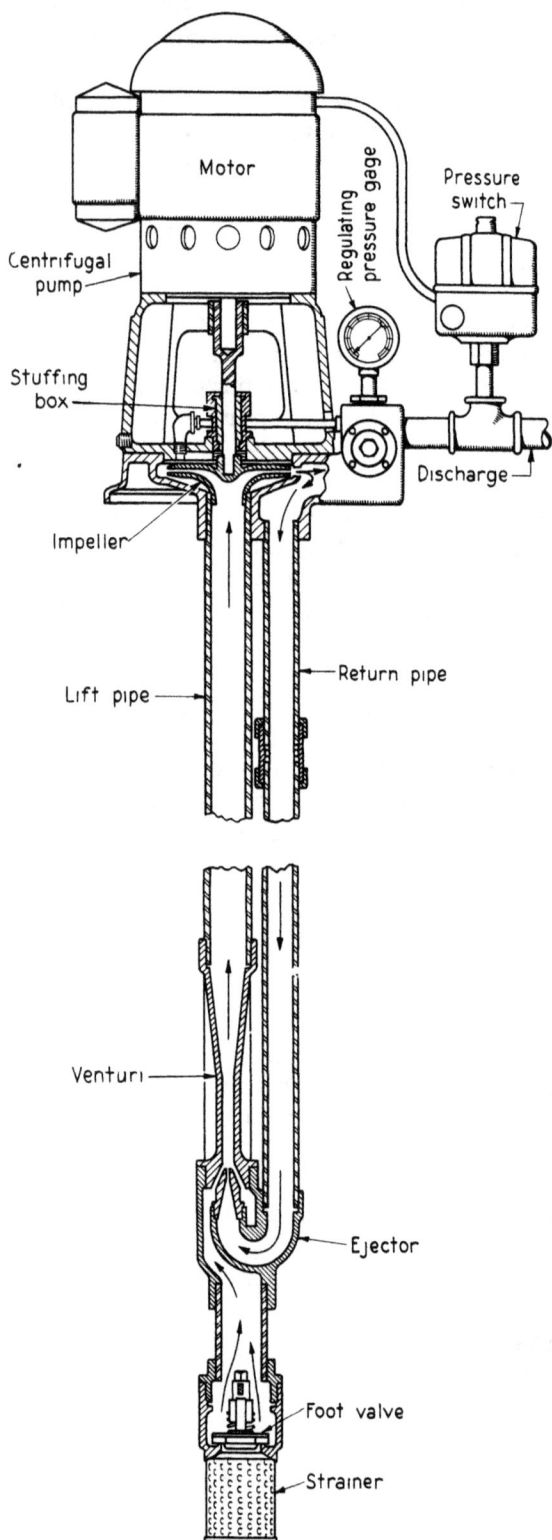

Figure 19. Ejector Type Pump

Deep well pumps are designed to lift water from depths greater than 22 feet. Of course the same type could be used for shallow wells, but because of a comparatively high first cost, their use is confined to deep wells.

Ejector Type

Ejector Pump

The ejector pump is relatively new to farm users. It is simple and of high capacity under low pressure conditions on the discharge side and to depths of 50 to 80 feet. Manufacturers do not generally recommend it for greater than 120-foot lifts. In fact, some recommend a maximum of 60- to 80-foot lift.

Pumping Mechanism. The pumping mechanism usually consists of a vertical centrifugal pump or a turbine pump of the type just discussed, operated in connection with a jet in the well. A reciprocating pump can be used but has not proved as popular as the other types mentioned. The centrifugal type is discussed here.

Water End. The ejector pump is really two pumps working together. The centrifugal pump at the surface lifts water about 15 feet by suction. At the discharge a part of the water is forced into the pressure tank line and a part down the well, through a return pipe, to the ejector.

The ejector acts as a second pump, creating a suction of its own, causing water from the well to pass through the foot valve and into the high velocity stream of the ejector. Water picked up from the well, plus water from the ejector, passes into the cone-shaped Venturi tube where the ejector continues to apply an upward lifting force. The ejector's job is to push the water up the lift pipe to where the centrifugal pump can lift it by suction and force it out the discharge.

The amount of water returned to the ejector increases with increased lift. Roughly, this represents about one-half of the total water pumped at a 50-foot lift and about three-fourths at a 100-foot lift.

Advantages of the ejector pump are:

1. It does not have to be installed over the well.
2. Pumping capacity increases as pressure lowers.
3. Pumping capacity increases as the water level rises in the well.
4. It is simple.
5. The pump itself requires no lubrication.

Disadvantages are:

1. Air in the suction or return pipe will completely stop water delivery from the pump.
2. Pumping capacity decreases rapidly with increased lift.
3. Some centrifugal systems have little or no provision for pumping air.
4. It cannot be used where high pressure is required. (Note table V.

When a turbine or piston pump is used, it is of the standard type with the addition of a regulator and the necessary fittings to provide for a return pressure pipe to the well.

If a centrifugal pump is used, it is usually of the vertical type with a bronze impeller driven by a stainless steel shaft. On the lower side of the impeller, and pressed into the suction line, is a metallic wearing ring

A bronze stuffing box, usually provided with metallic packing, fits around the stainless steel shaft. The pump frame and base are of cast iron.

The Venturi tube is usually of very accurately machined bronze and is sometimes lined with lead, porcelain, or rubber. The jet nozzle is usually of bronze or plastic. Below the jet is a brass or bronze foot valve.

Since jet pumps must have two pipes entering the well, the diameter of the casing is often a limiting factor. When the pipes run separately the casing size may be a minimum of 3 to 4 inches. A number of manufacturers are offering an assembly which provides for the lift pipe to fit inside the return pipe or vice versa. It is more expensive, but provides the advantage of greater water delivery than the twin types from the same size well casing.

Driving Mechanism. The driving mechanism of the vertical centrifugal pump is an electric motor with its shaft connected, by means of a coupling, to the stainless steel pump shaft.

Piston and Cylinder Type

Hand Operated Lift Pump

The hand operated lift pump for deep wells is comparable to the pitcher pump for shallow wells. The operating principles of the cylinder are exactly the same. The only fundamental difference in the pump design is in the placement of the cylinder. Instead of being a part of the pump standard, it is lowered into the well to within 22 feet of the water level or, better still, it may be submerged.

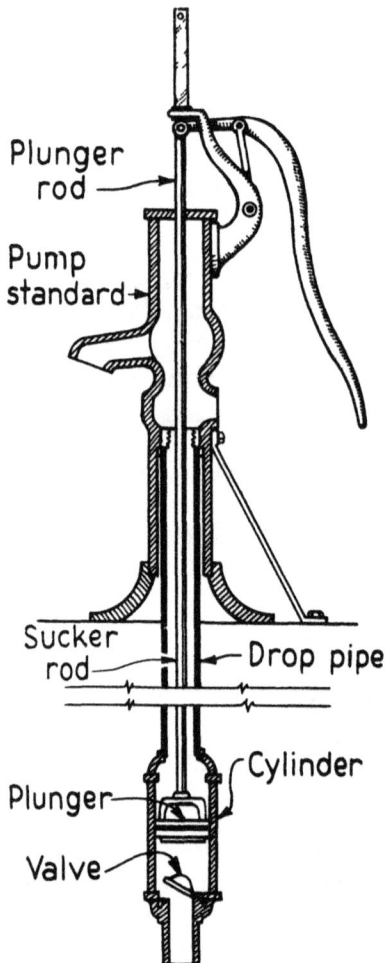

Figure 20. Common Hand Operated Lift Pump

The drop pipe holds the cylinder in position and at the same time acts as a delivery pipe between the cylinder and the pump standard. The pump plunger gets its motion through a sucker rod, which is connected at the upper end with the pump handle.

The action of the plunger on the water in the drop pipe is a straight lifting action independent of suction. If there is considerable distance from ground level to water level, the column of water in the drop pipe weighs heavily on the plunger and results in the pump being hard to operate.

Hand Operated Deep Well Force Pump

There are a number of designs for the hand operated deep well force pump, but one of the most common is shown in figure 21. It is very similar in appearance to the lift pump, except that it is designed to give the plunger rod a straight vertical motion to accommodate a stuffing box, and to provide for the attachment of a windmill or pump jack. The spout is usually provided with cock spout or shut-off valve.

The space between the delivery outlet and the stuffing box becomes an air chamber.

It is possible to make this type of pump automatic when used with a windmill. If a pump jack
is used with a gasoline engine or an electric
motor, it can be made semiautomatic with the
former and fully automatic with the latter by
means of the accessories discussed later.

For effective operation the pump should operate
at a speed of about 35 to 40 strokes per minute.
When power driven it is often operated much
faster resulting in excessive wear on the pump.
If the direction of rotation is not indicated
on the pump jack, it is well to check its operation to make certain the lift stroke comes
when the side arms are parallel with the plunger rod. Otherwise the side arms may apply
power at an angle with the pump and cause excessive wear on the plunger rod and guide.

A convenient and inexpensive arrangement for
connecting a motor and pump jack to a deep-
well force pump is shown in figure 105 of the
appendix.

Power Driven Deep Well Pump

Power driven hand pumps readily fall into
the deep well pump classification and fill
a very definite need on farms where the
pump must be operated by hand part of the
time. However, if a pump is to be operated
entirely with a gasoline engine or electric
motor, the specially designed driving mechanism commonly known as the
"power head" is better adapted. Several manufacturers make power heads
so that a handle can be added and the pump operated by hand, but the
majority of power heads installed on farms are for a power drive only.

Figure 21. Common Deep
Well Force Pump Used for
Hand, Windmill, and Pump
Jack Operation

Since the power head is generally used for automatic operation, we shall
take time to study its construction and the details of the pumping mechanism.

Water End. The operating principle in pumping water with a power head
is essentially the same as with a hand pump, except for the valves being
designed for quick action, additional leathers on the plunger, and certain provisions made to even the flow at the discharge.

The cylinder is installed in the well, preferably below water level, the
same as a hand pump. Two factors determine the size of cylinder to be
used:

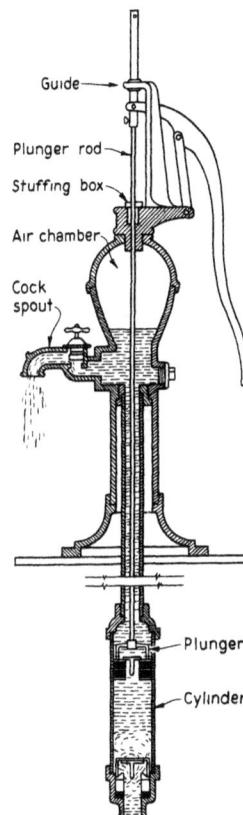

1. The diameter of the well casing or pipe in which the cylinder is to be installed.

2. The quantity of water the pump is expected to deliver per hour.

These two factors vary greatly with different installations, resulting in the manufacture of cylinders ranging in diameter from 1-13/16 inches to 5-3/4 inches, and in lengths to accommodate strokes of 6 to 24 inches.

Besides varying in dimensions, cylinders are constructed of different materials, such as iron, solid brass, or iron with a brass lining. Brass is most expensive, but for power operation either a brass or a brass-lined cylinder should be used.

Cylinders are of two common types known as open and closed. Most of us are acquainted with the closed type cylinder because of its use with deep-well hand pumps (figure 22a). The drop pipe between the pump standard and cylinder is smaller than the cylinder, so that in order to put new leathers on the plunger or repair the check valve the complete drop pipe and cylinder assembly must be drawn from the well. This necessitates considerable labor, as well as a risk of accidentally dropping the whole assembly into the well.

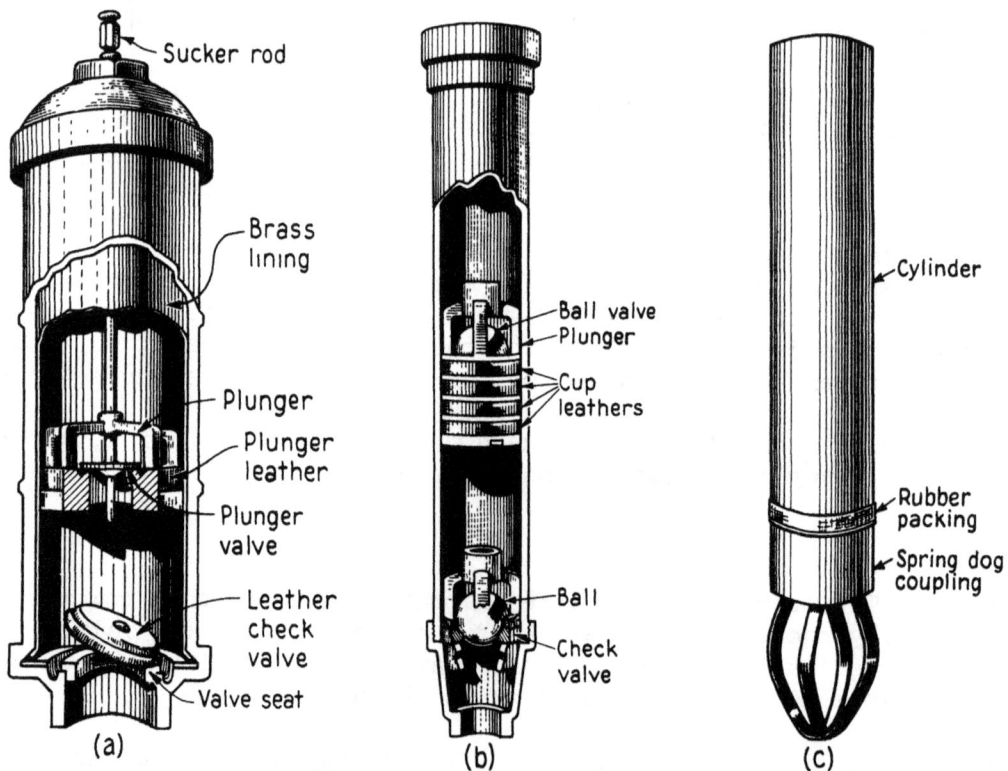

Figure 22. (a) Closed Type Cylinder Commonly Used on Hand Pumps
(b) Open Type Cylinder Popular in Many Sections on Power Pumps
(c) Tubular Well Cylinder

The <u>open</u> type consists of a cylinder of smaller diameter than the drop pipe (figure 22b). To remove the plunger for repairs the pump head is disconnected and the plunger pulled up through the drop pipe by means of the sucker rod. It is also possible to remove the check valve by pushing the plunger down and screwing it on the stem of the check valve and pulling them both up together.

The check valve is held in place by being wedged in the base of the cylinder, so that it need not be unscrewed for removal.

The open type installation is more expensive because larger drop pipe is used, but it is meeting with general favor and is usually recommended for most farm power installations.

It is not impossible to use the old hand pump cylinder for a power pump, but it is usually impractical unless it (1) shows little wear; (2) is of brass or is brass-lined; (3) has ample pumping capacity; and (4) is equipped with good valves.

In some sections the tubular well cylinder is popular. It is so designed that when lowered to the desired position in the well by means of a special seating tool attached to the drill rod, the cylinder shell can be revolved while the base remains stationary because of the gripping action of the spring dog coupling (figure 22c). When the cylinder is revolved, it expands a rubber packing which presses firmly against the well casing and forms an airtight and watertight seal.

The cylinder has special merit where it is desirable to have the largest possible size cylinder the well casing will accommodate. No drop pipe is required.

The <u>plunger</u> used in open and closed cylinders is the same. It consists of an iron or brass core (usually brass for power systems) with a valve in the center and one to four cup leathers contacting the cylinder. The greater the pumping load the more leathers used on the plunger.

<u>Double acting cylinders</u> are available for deep well pumps. They work on a somewhat different principle from that of the shallow well, double acting pump. They are rather complicated and are considerably more expensive than the cylinders just studied. Using the same length stroke and the same diameter cylinder, the double acting cylinder will deliver about 65 per cent more water than the single acting, which in some cases may well justify the additional cost.

The <u>plunger valve</u> may be any of the last four types listed below, except that glass valve seats are not used. The <u>check valve</u> may be any one of the five types listed.

1. Leather flapper valve with a brass or bronze seat (figure 22a).

2. Rubber faced poppet valve with a glass or brass seat (figure 23a)

3. Rubber faced spool valve with a brass seat (figure 23b).

4. Spring activated, rubber faced, poppet valve with a brass or glass seat (figure 23c).

5. Bronze ball valve on a brass seat (figure 22b).

The first type is limited to hand pumps.

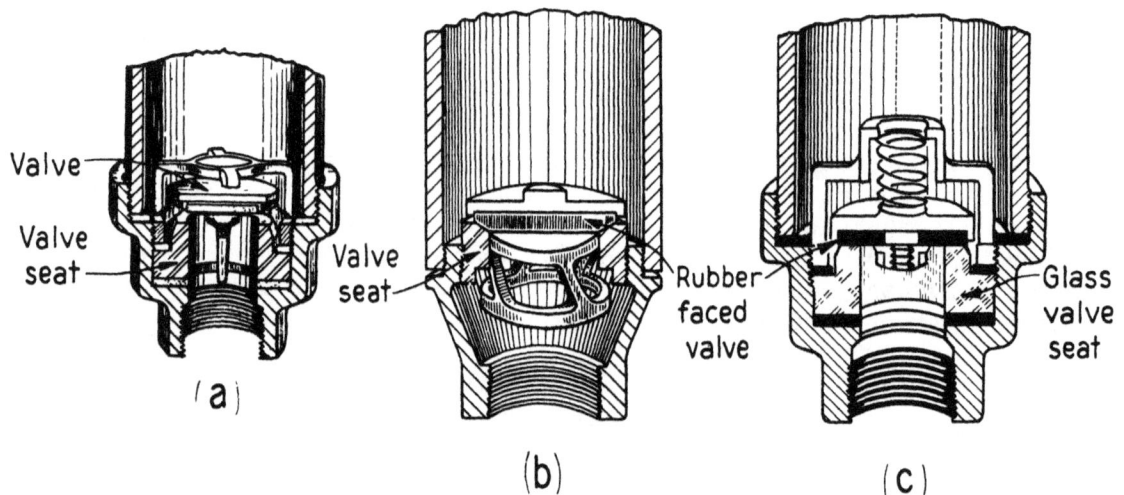

Figure 23. Valves Used with Deep Well Pump Cylinders
(a) Poppet Valve
(b) Spool Valve
(c) Spring Activated Poppet Valve

The sucker rod is the long rod that transmits power from the power head down through the drop pipe to the plunger in the cylinder. It may be made of wood or steel, but for power use wood is preferred because of its buoyance, quiet operation, and resistance to corrosion and shock. Since it does not corrode in water it will last much longer than steel.

Stuffing box and differential or pressure cylinder are terms applying to the part of the pump at or near the ground surface through which power is applied to the sucker rod.

The stuffing box works on exactly the same principle as that discussed for shallow well systems.

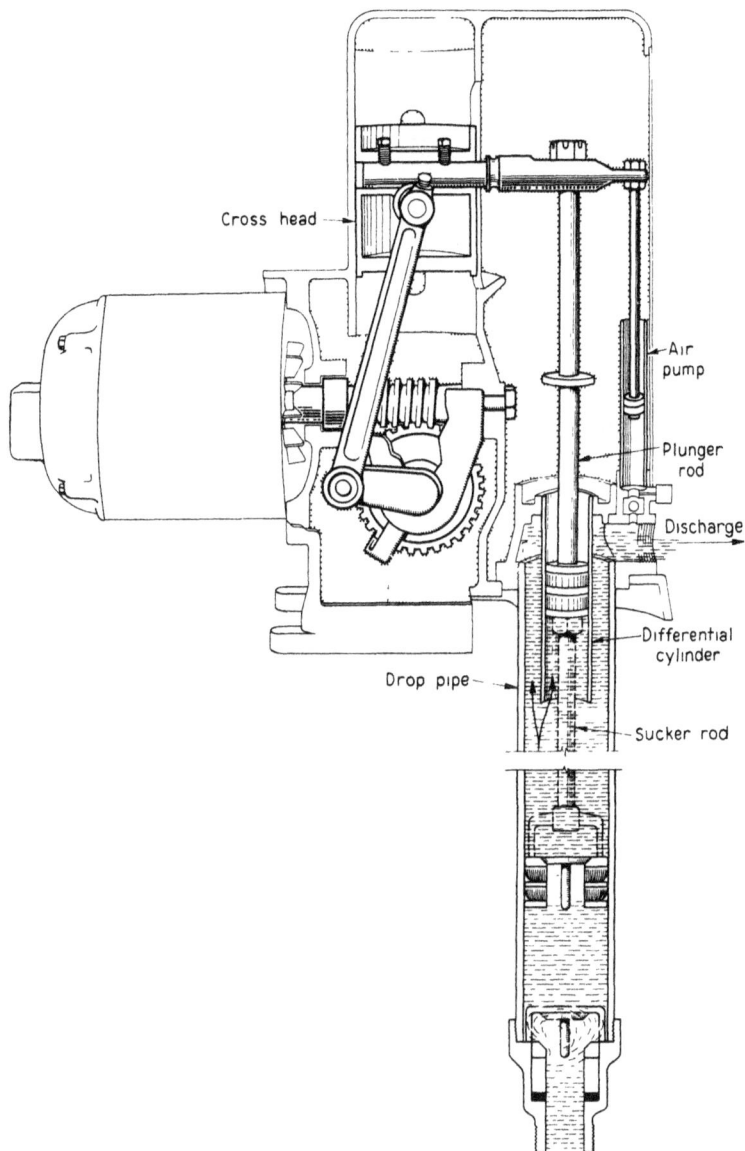

Figure 24. Differential Cylinder Replaces Stuffing Box,
Evens Flow at Discharge, and Balances Load
on Pump.

The differential cylinder is another cylinder, independent of the one in
the well, mounted on or near the power head and with a plunger that moves
up and down with the sucker rod. It accomplishes the job of a stuffing
box; also on the up stroke part or all of the water delivered up the drop
pipe enters the differential cylinder instead of the discharge pipe. On
the down stroke it is discharged. Since this arrangement enables the

pump to lift water on the up stroke and discharge it under pressure on t down stroke, the strain on the power head is more evenly distributed and the water discharge pulsates less.

Some manufacturers use a larger differential cylinder than well cylinder so that on the up stroke the differential cylinder has the greater displacement, which creates a partial vacuum and causes air to enter the cylinder. The reason for this will be discussed further under "Pressure Tanks."

<u>Driving Mechanism</u>. The driving mechanism of a deep well power head diffe from a power driven shallow well pump in three major respects:

1. It is more heavily constructed.

2. Vertical motion, instead of horizontal, is applied to the pump r

3. The plunger operates slower; consequently more provision is made for speed reduction.

The <u>power source</u> is usually a gasoline engine or an electric motor belted to the power head; however, there are a number of farm water systems buil with a direct drive as shown in figure 24.

The power head pulley on the belted pump is mounted on a shaft, which extends through the pump housing and supplies power to one or two small machine cut gears. These in turn supply power to the same number of large gears, which in turn supply power through a oonnecting rod to a crosshead or walking beam. The crosshead works the same way as with the shallow well power pumps, except that the motion is vertical (figure 25).

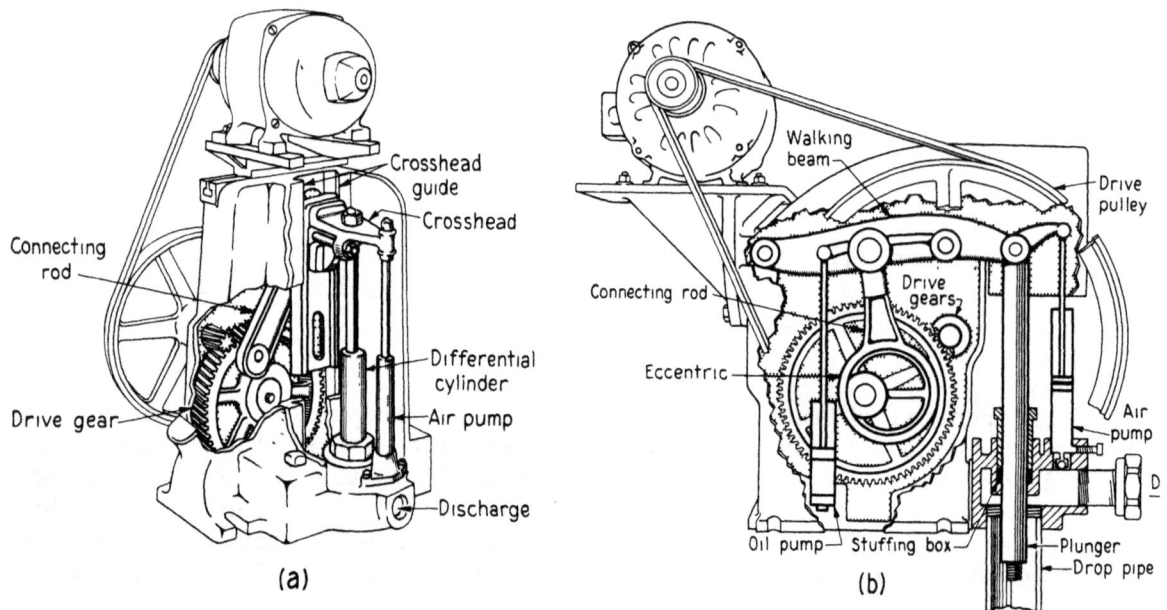

Figure 25. (a) Crosshead Type Deep Well Pump
            (b) Walking Beam Type

The walking beam usually consists of a horizontal lever with one end secured by means of a flexible link to the gear housing. Power is applied near the center and the plunger rod is attached at the other end (figure 25b).

Some manufacturers use a combination of both principles.

Low-cost pumps are available that eliminate gearing and supply power to the crosshead in the manner of a shallow well pump. Their use is limited to small capacity requirements (figure 26).

Lubrication is usually accomplished by means of one or more of the following: (1) separate oil pump; (2) pressure created by gear mesh; (3) revolving chain; (4) splash system; (5) revolving oil rings; (6) oil wicks; and (7) grease seal bearings.

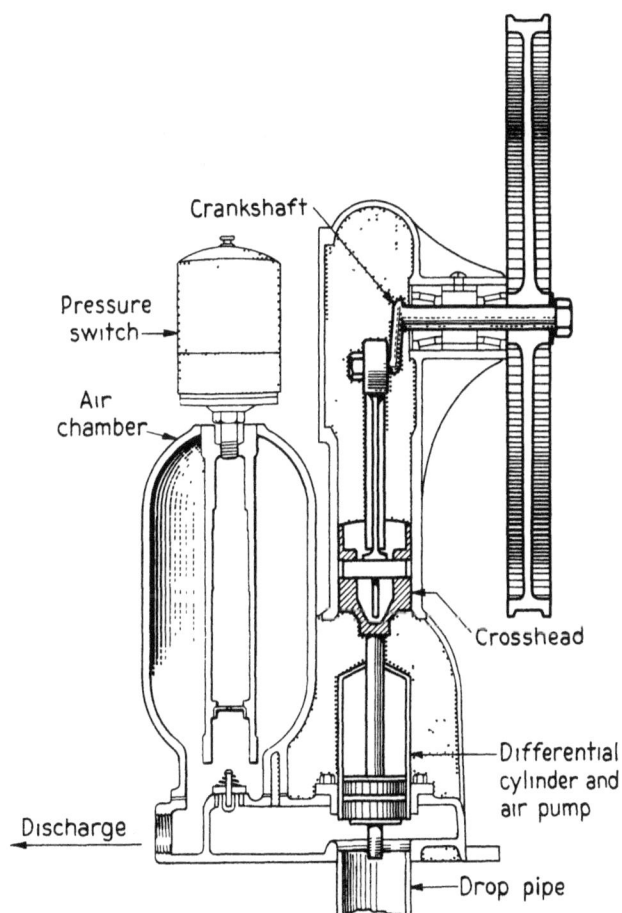

Figure 26. A Low Capacity Gearless Type Deep Well Pump

CHAPTER IV

WATER SYSTEM PRESSURE--PUMP ACCESSORIES

Pressure at the Faucet

Pressure, as most of us think of it, is the force with which water flows from an open faucet. There are three common ways in which pressure is secured for farm installations:

1. Direct mechanical action.
2. Gravity tank.
3. Hydro-pneumatic tanks.

Direct Mechanical Action

Direct mechanical action is the least common form of pressure and is normally not practical with power pumps unless fresh water is desired. The system is sometimes used to keep down installation costs, and can be used only with electric pumps or hydraulic rams, as will be discussed later.

With an electric pump no water storage is used. When the faucet is opened the pressure immediately drops, the pump starts and continues to pump until the faucet is closed. If the faucet does not discharge the water as fast as it is pumped, the motor may start and stop several times. If two faucets are open at the same time, neither will get sufficient pressure. The frequent starting and stopping is hard on both motor and pump.

The system is used satisfactorily in combination with a pressure tank when a special arrangement supplies fresh water to one or two faucets for drinking purposes, while the remaining outlets are supplied from the storage tank (figure 32).

Gravity Tank

The gravity tank has been a common means of securing pressure and is still used to some extent. It consists of an elevated wood, steel, or concrete tank. Concrete tanks are largely limited to mounting on hills while wood and steel tanks are commonly mounted in a building or on a tower. Wood tanks have the advantages of not sweating or rusting.

A gravity tank is desirable where large water storage is needed or where the power source is a windmill or gasoline engine. With most new systems, particularly electric, it is not used.

The principal disadvantages are that:

1. There is greater possibility of water contamination.
2. It is expensive to erect.
3. It is subject to freezing.
4. The water becomes warm in summer.
5. It may be somewhat unsightly.
6. There is low pressure at the faucet.

Low pressure will not exist if the tank is mounted sufficiently high. An elevation of 25 to 30 feet will give a pressure of 11 to 13 pounds per square inch. Most farm installations do not exceed that height.

If a gravity tank is mounted in the attic of a home or in the hay mow of a barn, as is often the case, there is a possibility of overtaxing the strength of the structure when the tank is filled. Consequently, it should rest directly on the weight-bearing part of the structure.

A disagreeable feature of metal attic tanks is "sweating" during warm weather. It can be taken care of, to some extent, by resting the tank on a metal pan which drains the condensed moisture to the exterior. This may also take care of nonautomatic systems when the tank overflows; however, every precaution should be taken to install an overflow pipe of sufficient capacity to avoid the possibility of the water reaching the top of the tank.

Hydro-pneumatic Tanks

Hydro-pneumatic tanks are commonly called "pressure tanks," and will be referred to as such in this discussion. They are completely enclosed except for pipe tappings and vary in size from 12 gallons to 3000 gallons capacity. The 42-gallon size is most common on electric farm water systems. Considerably larger tanks are used for hand or engine driven pumps. Either black or galvanized steel tanks are available, the latter giving longer service.

Pressure is secured by trapping the original air in the tank and forcing water in with it. Since water does not compress and air does, the air is forced to occupy less and less space, and in so doing exerts more and more pressure on the incoming water.

With an automatic pressure switch the pump is normally stopped at 40 pounds pressure, at which time the tank is about two-thirds full of water. With a 42-gallon tank, when about 7 gallons of water is drawn, the water level is a little below the halfway mark and the pressure has dropped to 20 pounds, causing the pump to start.

More water can be drawn with the remaining 20 pounds pressure in case of a power interruption.

A pressure tank has the following advantages:

1. It is relatively small.
2. It is inexpensive for small sizes.
3. It provides satisfactory pressure.
4. It is easily installed.
5. It can be installed in an inconspicuous place.

Its principal disadvantages are:

1. Its active storage capacity is generally not as great as that of a gravity tank.
2. It requires the frequent addition of more air.

The last point is often overlooked or not understood by farm people. The passage of water through the tank will gradually absorb the air and exhaust the supply. When this happens the tank loses its capacity to deliver water, and, instead of 7 gallons being drawn, the usual capacity may be reduced to a gallon or less before the pressure drops from 40 to 20 pounds. This condition is called "water logged," and is corrected by adding more air.

It is possible to add so much air that it will escape through the delivery pipe with the water and cause alternate spurts of . air and water.

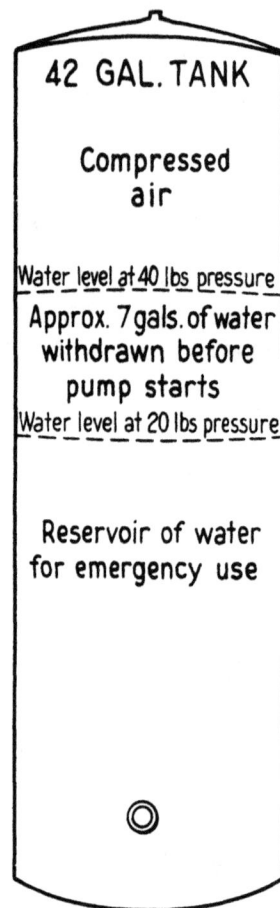

42 GAL. TANK

Compressed air

Water level at 40 lbs pressure
Approx. 7 gals. of water withdrawn before pump starts
Water level at 20 lbs pressure

Reservoir of water for emergency use

Figure 27. Hydro-pneumat Tank, Generally Known as Pressure Tank

<u>Addition of air to pressure tanks</u> is accomplished in several ways.

1. The tank may be completely drained and again filled with water.
2. Some tanks have provision for attaching an ordinary bicycle pump and forcing air in by hand.
3. Most automatic piston type shallow well pumps have provision for sucking in a small quantity of air through the water chamber. The air is delivered to the tank along with the water (figure 15).

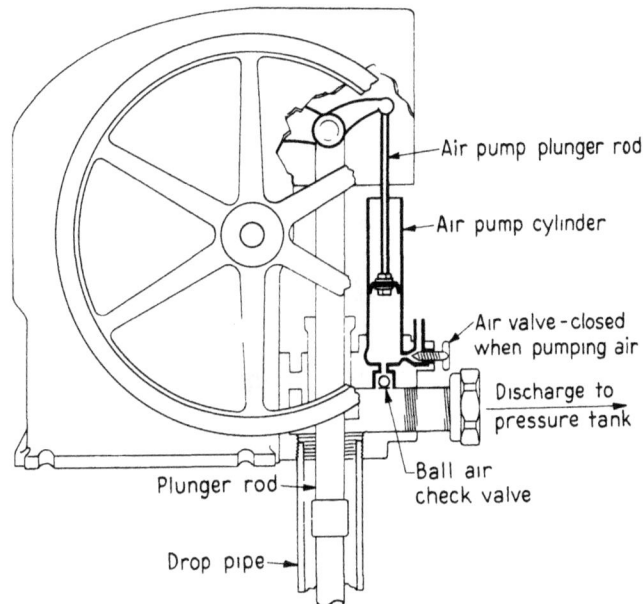

Figure 28. Air for Pressure Tank Is Supplied by
Separate Pump on Deep Well Head

4. Deep well piston pumps are commonly provided with a separate
   air pump operated in connection with crosshead or walking
   beam (figure 28).
5. Still other less common principles are:

   a. A diaphragm air pump on shallow well piston pump
      activated by alternate suction and pressure.
   b. Jet principle, where water passing a jet at high
      velocity creates a vacuum and air forces in.
   c. A diaphragm, or the equivalent, which is separately
      encased and connected on one side with the suction
      line and with the pressure tank on the other. When
      the pump starts, the diaphragm is drawn out of posi-
      tion and air enters through an outside valve. When
      the pump stops, a spring or back pressure forces the
      air into the pressure tank. In case the water level
      in the pressure tank is below the level of the open-
      ing to the diaphragm, air enters the diaphragm chamber
      from the tank rather than through the snifter valve.
      The device is used mostly with centrifugal pumps
      (figure 29).

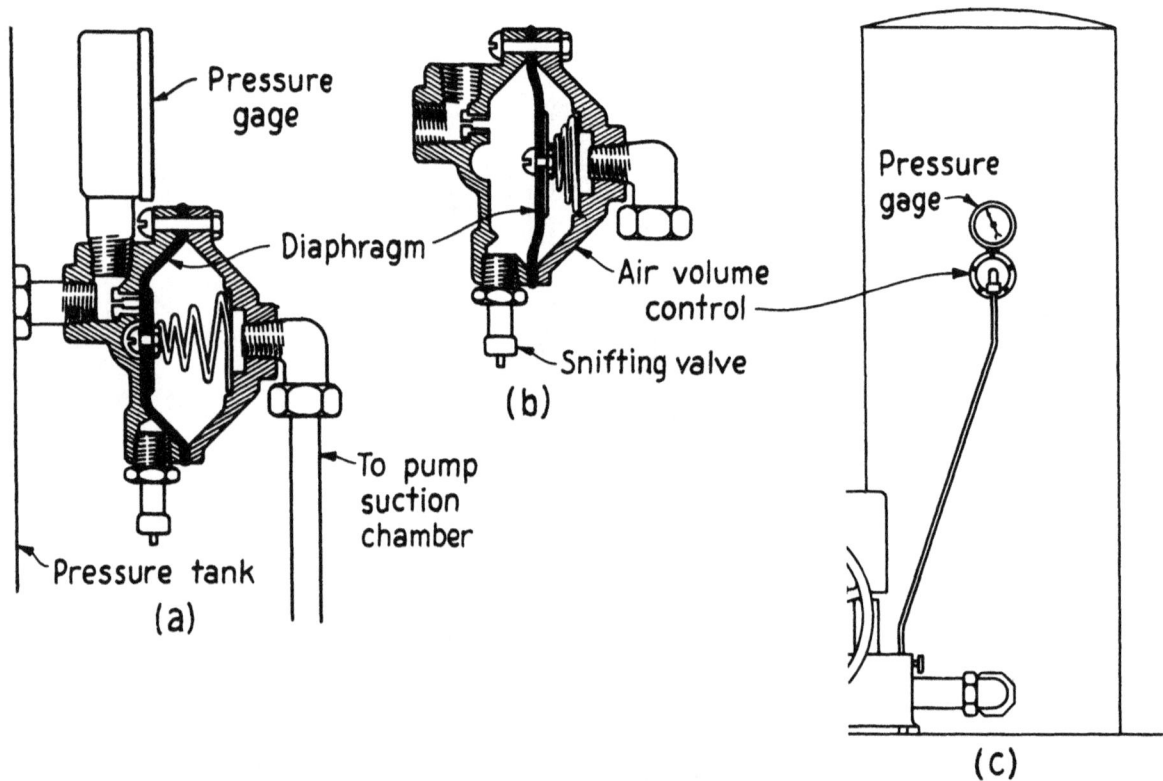

Figure 29. Diaphragm Type Air Volume Control
(a) Diaphragm Position When Pump
Is Not Operating
(b) Diaphragm Position During Operation
(c) Mounting on Pressure Tank

Automatic Air Control. Most farm water systems on the market are equipped with automatic air control of some form.

To make a piston turbine or rotary type shallow well water system automatic it is necessary to extend a tube from the air inlet on the pump to a float controlled valve on the pressure tank. The float follows the water level inside the tank and connects with the valve on the outside by means of an arm. When the water and air are in correct proportion the valve remains closed. But as quickly as the air supply commences to get low the water level rises, forcing the float higher and causing the valve to open and permit outside air to pass down through the tube where it can be drawn by the pump into the cylinder and discharged into the pressure tank with the water (figure 30a).

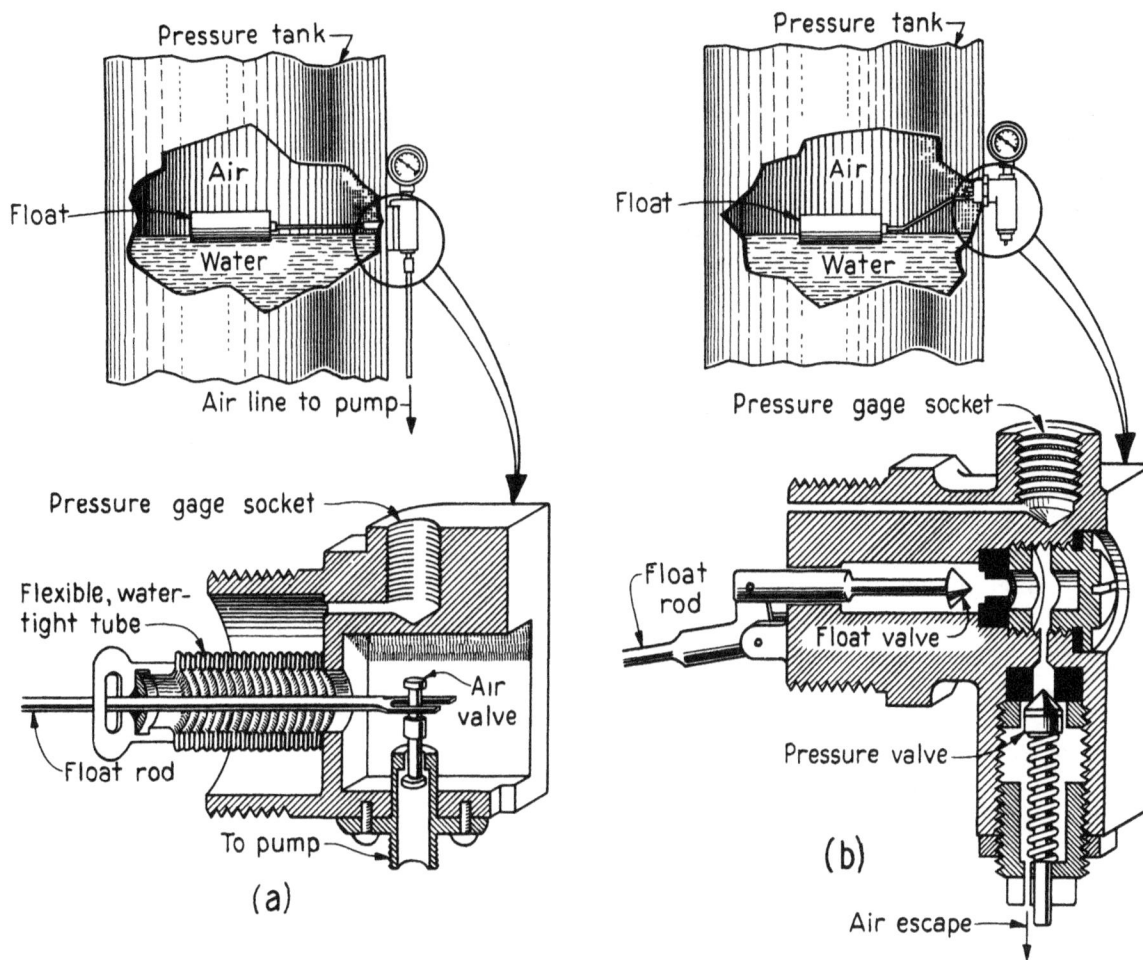

Figure 30. Pressure Tank Air Control Valves Used
with Piston Pumps
(a) Shallow Well Type
(b) Deep Well Type

The deep well arrangement is different. Each time the pump operates, a small amount of air is forced in with the water. When too much air collects in the tank the float mechanism opens a valve and a portion of the air is released (figure 30b).

Accessories

The pieces of equipment to be discussed are listed as accessories because they can be purchased as separate units and used to modernize many existing water systems. Most automatic water systems sold now have the necessary accessories included.

Pressure Switch

Automatic switches are particularly well adapted to electric power, but they can also be used to break the ignition circuit on a gasoline engine driven pump. If used with an electric motor and pressure tank, the switch is normally set to start the motor when the pressure drops to about 20 pounds per square inch and to stop it when the pressure rises to 40 pounds per square inch.

The mechanism varies with different makes of switches, but in general it consists of a spring and a diaphragm which is exposed to tank pressure. As the pressure drops the spring forces the diaphragm to recede. The diaphragm is connected with a lever or system of levers which close the small electric switch and start the motor. When the water pressure rises the diaphragm action is opposite, causing the levers to open the switch and stop the motor.

Nearly all pressure switches have some provision for <u>adjustment</u>, so that operating pressures may be increased or decreased. The necessity for

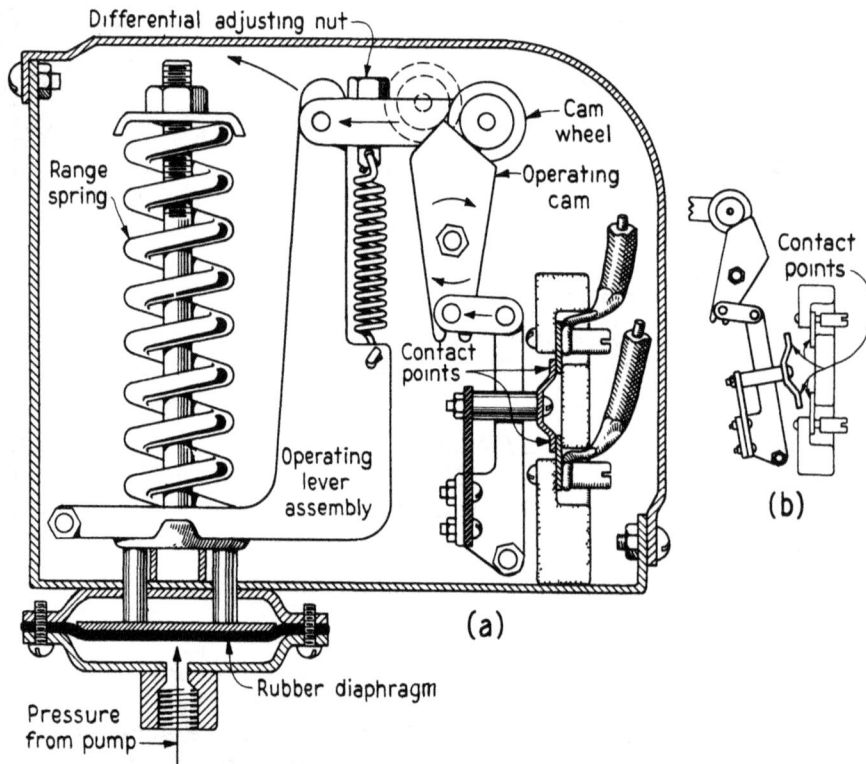

Figure 31. One Type of Pressure Switch
(a) Electrical Circuit Closed
(b) Circuit Open

using higher pressure will be discussed in chapter V. It is not advisable for the average person not acquainted with pressure switches to attempt to adjust them.

Some switches are designed to open only one side of a two-wire circuit and are known as single pole switches. They should be used for motors connected to 115-volt circuits. If the switch breaks both sides of a two-wire circuit, it is a double pole switch and can be used on either 115- or 230-volt circuits. Either type can be used to open the ignition circuit of a gasoline engine driven pump.

Some pressure switches are provided with overload protection for an electric motor. It consists of a small heating device which trips when the motor is overloaded and opens the circuit.

Float Switch

When water is pumped into an elevated storage tank a float type switch can be used. The float is adjustable so that it will start an electric motor driven pump when the water level gets low and stop it when the water has reached the desired level. The same switch can also be used to stop a gasoline engine.

Check Valve

A check valve is designed to permit water passage one way only (figure 32) It is desirable with pressure pumps to relieve the pump from back pressure when it is not in operation; consequently a check valve is installed on the discharge line close to the pump.

Air Chamber

An air chamber may be used on piston pumps on the discharge line as shown in figure 32. Its purpose is to provide an air cushion to absorb the shock of alternate starting and stopping of the water through the pipes while the pump is operating.

Vacuum Chamber

A vacuum chamber accomplishes the same thing on the suction line that the air chamber does on the discharge line. However, instead of containing air under pressure, it consists of a partial vacuum. Some pumps have both a vacuum and air chamber built on.

Relief Valve

A relief valve is standard with pressure pumps. It is simply a valve held in place by spring pressure, which is set for greater than normal pump operating pressure. If for some reason pressure builds up in the system beyond normal, the relief valve opens and permits enough water to escape to lower the pressure (figure 32).

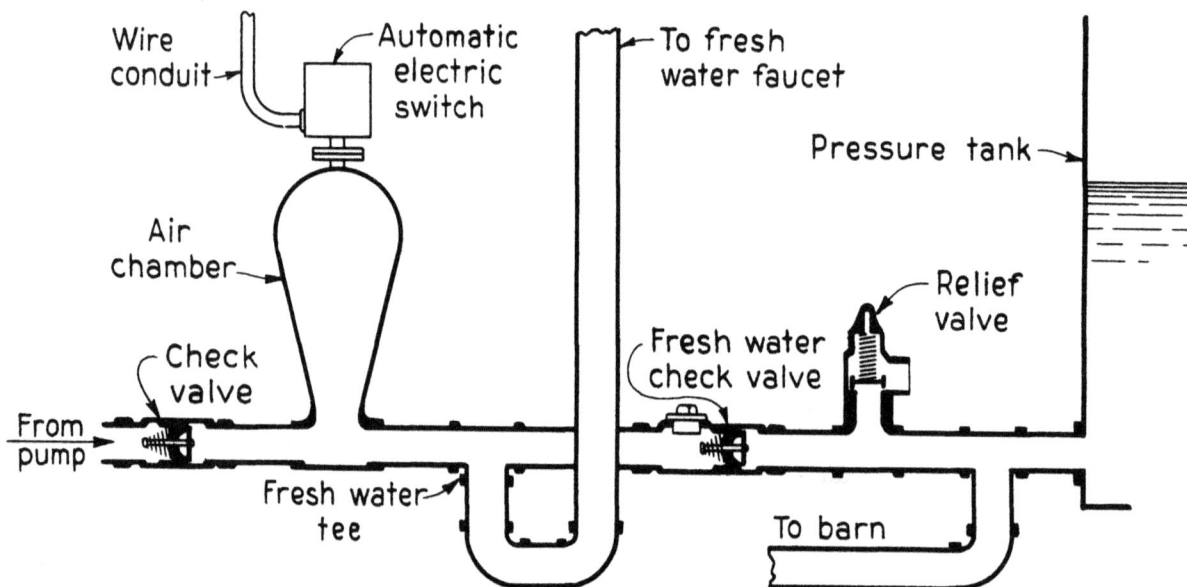

Figure 32. Arrangement Used with an Automatic Pump and Pressure Tank to Supply Fresh Water Directly from Well

Fresh Water Valve

A fresh water valve may be used with an automatic pump and pressure tank. It provides for water to be pumped directly from the well without passing through the tank, thus making it more desirable for drinking purposes. It consists of another check valve, and connects to the discharge line in the manner indicated (figure 32). When the fresh water faucet is opened the pressure immediately drops, causing the automatic switch to start the motor and pump. When the faucet is closed the pump stops unless the tank pressure is below 40 pounds; in that case it will continue to pump until that pressure is reached.

A small leak in the fresh water line will cause the pump to operate frequently, and for only a fraction of a minute. To avoid this some provision is usually made in the fresh water check valve to permit a small amount of water to leak back from the pressure tank. The fresh water tee must point down to keep the pipe from filling with air; even then, some air which is entrained with the water will accumulate at the faucet.

The desirability of the arrangement is sometimes overstressed. If water consumption is sufficient to change the contents of the tank several times daily, the water in the tank is seldom more than two or three degrees warmer than in the well.

Pressure Gage

A pressure gage is usually used to indicate the existing pounds pressure in a pneumatic tank. It is not standard equipment with all pressure pumps, but it is sometimes valuable in detecting pump trouble and is essential if a pressure switch setting is to be changed to operate at higher or lower pressures. On jet pumps a pressure gage is used at the pump discharge for regulation purposes.

Water Gage

Since the acceptance of automatic air control on water systems, the water gage has little value. It was formerly used to check on the water level in pressure tanks, so that when the air supply became low the air valve on the pump could be opened to enter air and closed again when the water level receded sufficiently (figure 33).

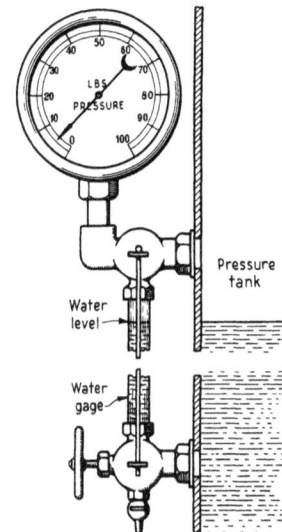

Figure 33. Water Gage and Pressure Gage for Pressure Tank

Motors

Three types of electric motors are in use on electric water systems. They are (1) split phase, (2) repulsion-induction, and (3) condensor-start types. The first is least costly but is not generally satisfactory. It requires a heavy current to start, causing low voltage on the circuit. If the pump is very hard to start, the starting windings may burn out after a short period of operation.

The last two types of motors are satisfactory.

If electricity is not available belt-driven pumps can be operated with gasoline power. Small high speed engines are commonly furnished which can be readily replaced with an electric motor at a later date since both operate at about the same speed.

CHAPTER V

PUMP SELECTION AND INSTALLATION

Discussions in this chapter are confined to the selection and installation of automatic and semiautomatic water systems. However, certain of the information is quite applicable to hand operated pumps.

Pump and Tank Selection

Farm Water Requirements

As was mentioned in chapter I, it is very important that the source of water supply be adequate, dependable, and safe. If these conditions do not exist, the installation of a water system is of questionable value.

A common mistake is to install a pump with insufficient capacity. Most farm families think of a water system in terms of the amount of water they carry by hand or pump with a hand pump, so that a system rated at 120 gallons per hour, for example, seems quite adequate. In some cases it may be sufficient, but most farm families will outgrow the capacity of that size pump in a few years.

There is, however, a limiting factor of some wells to supply water in the quantities desired and often there is little that can be done about it. If a dug well is the present source and considerable expense is required to cover it and make it safe, it would be well to investigate the cost of a drilled well. In some sections a well could be drilled at little more cost than that of improving the old dug well, and the water supply would be more likely to be adequate and dependable.

A pump should not be installed that is of greater capacity than the well If adequate water is available, the probable consumption and capacity pump needed can be figured from tables II and III.

Consumptions indicated in table II, as compared with 6 gallons per person a day when water is carried by hand, give a good clue as to why few farm water systems are large enough.

TABLE II

Water Consumption

For each member of family (total average daily
  consumption allowing for kitchen, bathroom,
  laundry and some sprinkling) . . . . 35 gallons a day
*For each cow . . . . 30 gallons a day
For each horse (winter, 4 to 8 gallons;
  summer, 8 to 18 gallons) . . . . 15 gallons a day
For each hog . . . . 2 gallons a day
For each sheep . . . . 1-1/2 gallons a day
For each 100 chickens . . . . 2-1/2 gallons a day
For sprinkling (1/2-inch hose) . . . . 200 gallons an hour
For sprinkling (3/4-inch hose) . . . . 275-300 gallons an hour
  (10 gallons will sprinkle 100 sq ft;
  20 gallons will soak 100 sq ft)

*Statistics show that high producing cows sometimes drink as much as
35 to 40 gallons a day.

Figuring Pump and Tank Sizes

In order to arrive at the pump and pressure tank requirements in a simple
manner at least one manufacturer has adopted the plan shown in the following table:

TABLE III

Water Requirements as Related to
Pump and Tank Capacities

	Pump Gal. per Hr.	Tank Cap. Gal.
Minimum fixture flow (always added)	125	
Additional Requirements		
Kitchen sink	75	5
Bathroom, water closet		5
Bathroom, lavatory		5
Bathroom, tub		10
Basement, water closet		5
Bathroom, shower bath	100	
Basement, laundry tray		10
1/2-inch Garden sprinkling hose	75	
3/4-inch Garden sprinkling hose	100	
Automatic lawn sprinkler	125	
Livestock, per head	5	1

To determine pump capacity needed it is only necessary to select the various water uses a farm family expects to have and under the heading of "Pump" total the figures opposite the anticipated uses. The total will be the minimum capacity pump needed.

The same procedure is followed in the column headed "Tank." However, if the tank capacity should figure considerably less than 42 gallons, it is well in most cases to use a 42-gallon tank as the minimum.

If a gravity tank is to be used, the pump capacity can still be the same but the tank capacity would be larger, since a gravity tank is usually installed where it is desirable to store large quantities of water.

Another method of figuring pump capacity is to take the first table and estimate the farm needs. After all the needs are totaled divide by 2 to get the pump capacity. This method is based on the idea that a pump should be large enough that two hours of operation per day are sufficient.

For fire protection it is well to figure pump capacity at 8 to 10 gallons per minute, which is sufficient to supply two hose lines. That quantity of water will not extinguish a fire that is well under way, but it is very effective on a small fire or for protecting nearby buildings when there is a large fire. At present, insurance companies make no reduction in premiums when automatic farm water systems are installed.

Pipe Friction

Pipe friction is discussed at this time because of its relation to pump installations. However, the pipe friction problem is present wherever water passes through a pipe, whether from a spring on a hillside, hydraulic ram, or pump. Friction is caused by the inside surface of a pipe retarding the flow of water next to it, which in turn tends to retard the whole flow (figure 34).

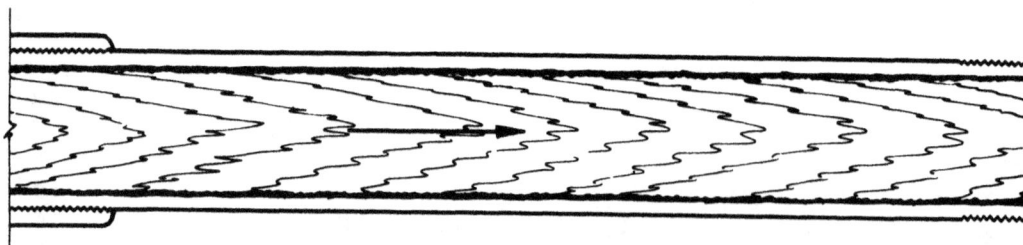

Figure 34. Water Flow Is Retarded Next to the Pipe.

It has been found that friction increases as:

1. Rate of flow increases.
2. Pipe length increases.
3. Pipe diameter decreases.
4. Pipe roughness increases.
5. Bends are added.

To illustrate how effective pipe friction can be, note figure 35. In 35a a column of water 1 foot high exerts enough pressure to cause a flow of 20 gallons per minute through 400 feet of 3-inch pipe. By simply reducing the pipe size to 2-inch, as in 35b, it takes a column of water 7.28 feet high to create enough pressure for a 20-gallon per minute flow. If a pump were supplying the pressure instead of the water column, we can realize it would require less than one-seventh of the power to overcome friction in a 3-inch pipe as compared with the 2-inch.

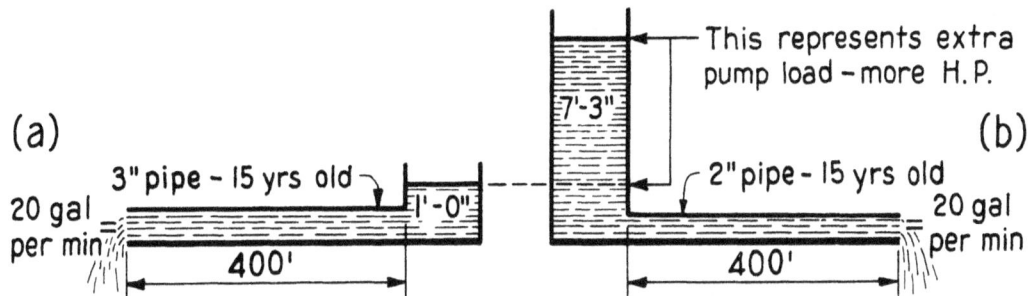

Figure 35. For Equal Flow a Small Pipe Requires
More Pressure than a Large One.

You will note from the illustration that it is possible to measure friction by the height water column necessary to overcome it.

Friction can also be measured in pounds pressure necessary to overcome it. Both are shown in the following table.

TABLE IV

Pipe Friction per 100 Feet
Ordinary Iron Pipe
15 Years Old

Flow Gals. per Min.	Size of Pipe											
	1/2 inch		3/4 inch		1 inch		1-1/4 inch		1-1/2 inch		2 inch	
	Ft.	Lbs.	Ft.	Lbs.	Ft.	Lbs.	Ft.	Lbs.	Ft.	Lbs.	Ft.	Lbs.
2	7.4	3.2	1.9	.82								
3	15.8	6.85	4.1	1.78	1.26	.55						
4	27.0	11.7	7.0	3.04	2.14	.93	.57	.25	.26	.11		
5	41.0	17.8	10.5	4.56	3.25	1.41	.84	.36	.40	.17		
6			14.7	6.36	4.55	1.97	1.20	.52	.56	.24	.20	.086
8			25.0	10.8	7.8	3.38	2.03	.88	.95	.41	.33	.143
10			38.0	16.4	11.7	5.07	3.05	1.32	1.43	.62	.50	.216
12					16.4	7.10	4.3	1.86	2.01	.87	.70	.303
14					22.0	9.52	5.7	2.46	2.68	1.16	.94	.406
16					28.0	12.10	7.3	3.16	3.41	1.47	1.20	.520
18							9.1	3.94	4.24	1.83	1.49	.645

Feet of pipe equivalent to a 90-degree elbow

5	6	6	8	8	8

In chapter II we learned that an atmospheric pressure of 14.7 pounds per square inch would raise a column of water 33.9 feet. The same ratio still holds. In other words

1 lb. pressure = 2.31 ft. water column
1 ft. water column = .43 lb. pressure

The same ratio holds throughout the above table.

Suction Lift

Figuring Suction Lift

Suction lift is made up of two items:

1. Vertical distance from pump level to the lowest probable water level in the well.
2. Friction loss in the suction pipe.

For shallow well pumps we learned that 22 feet is considered the maximum practical suction lift. If the water level in a well draws down much below that depth, the only recourse is to install a deep well pump.

When pipe friction is figured, it may develop that a shallow well pump will not lift from a depth of 22 feet because atmospheric pressure cannot exert enough force to lift water 22 feet and at the same time overcome the pipe friction.

From a practical standpoint, if a pump is mounted at the well and a suction pipe is used that is no smaller than the suction tap on the pump, most pump men neglect friction and figure an effective pumping depth of 22 feet.

The situation is considerably different, though, in the case of a pump mounted some distance from the well.

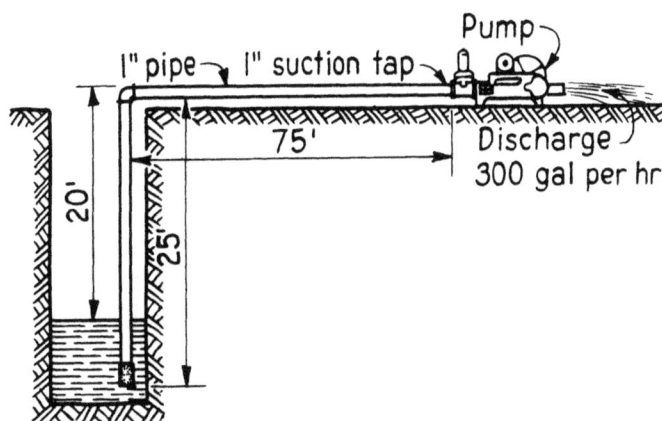

Figure 36. Pipe Friction Should Be Considered When
a Pump Is Distant from the Well.

Note the conditions under which the pump works in figure 36. Whether or not the pump is likely to work satisfactorily can be figured as follows:

Height water is lifted =   20 feet
Pipe friction:

Pipe in well	25 feet	
Pipe from well to pump	75 feet	
1-inch elbow = in straight pipe	6 feet	
Total	106 feet	

Friction loss in 100 feet of 1-inch pipe at 5 gallons
  per minute flow = 3.25 feet per 100 feet of pipe
  (table IV)
Since there are 106 feet, friction loss figures
  (106 + 100) x 3.25 =   3.4 feet
                  Total lift =   23.4 feet

46

It is possible that such an installation will work, but it is not usually
safe to assume that it will. To correct the situation, if 1-1/4-inch
suction pipe is used (table IV), the friction is reduced to .9 feet, making a total suction lift of 20.9 feet, which is within the practical suction limit.

Manufacturers of jet pumps provide tables which show the necessary pipe
size for different installations when the pump is mounted over the well.
In case a jet pump is mounted some distance from the well, pipe friction
must again be figured or the efficiency of the pump may be decreased.
Centrifugal and turbine pumps, usually used with jet pumps, decrease in
efficiency by delivering less water, not by increasing horsepower demands

Likewise, in the installation of deep well piston type pumps, tables are
furnished by manufacturers. Since this type of pump must be set over
the well, there is no problem of horizontal suction. Vertical suction
is often reduced to a minimum because of the cylinder being placed below
water level.

Discharge Head

If a pump is required to lift water to its own level and discharge into
open air, there is no problem of a discharge head. However, any pump,
whether hand or power operated, that delivers water to a gravity tank,
to a pressure tank, or through a length of pipe, is working under added
difficulties.

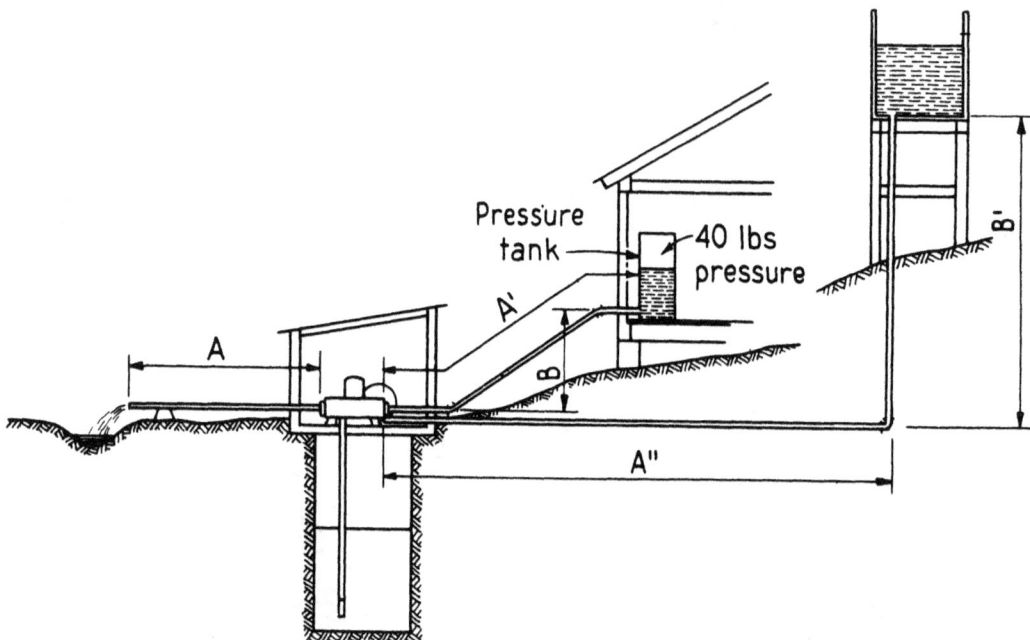

Figure 37. Three Common Pressure Head Conditions

Any one, two, or three of the following constitute discharge head:

    1. Vertical distance from pump to outlet.
    2. Friction loss in discharge line (friction head).
    3. Pressure required at discharge outlet.

Example (1).   To illustrate, assume that the pump in figure 37 is pumping into the open ditch at the left of the pump house. Pipe size is 1 inch, the distance A 100 feet, and the flow 5 gallons per minute. From table IV we find the <u>discharge head</u> is 3.25 feet or, in other words, it would take a column of water 3.25 feet high to supply enough pressure to overcome the resistance in the 100 feet of pipe. The 3.25 foot column of water would exert a pressure of 1.41 pounds (table IV), which is the extra pressure the pump must develop.

Example (2).   Assume that the pump is pumping into the gravity tank on the right. Besides pipe friction we have the additional task of lifting water the vertical height B' plus a 90-degree elbow in the line.

Assume a flow of 5 gallons per minute through 1-inch pipe, with A" distance 100 feet and B' distance 25 feet. The discharge head figures as follows:

```
Vertical lift B' =                                    25 feet
125' + 6' = 131' on which to figure pipe
   friction = (131 + 100) x 3.25 =                   4.25 feet
                              Discharge head      29.25 feet
```

To get the equivalent in pounds pressure divide 29.25 by 2.31 = 12.6 pounds discharge head.

Example (3).   Assume that the same pump is delivering 5 gallons per minute into the pressure tank through 1-inch pipe. If A' is 100 feet and B is 5 feet, it would figure as follows:

```
Pipe friction A' 100 foot length plus 12 feet for
   two elbows = 112 feet = Friction head
   (table IV) of                                      3.64 feet
Vertical height B =                                    5.00 feet
                                     Total =           8.64 feet
```

8.64 + 2.31 = 3.7 pounds pressure head in getting water to pressure tank.

3.7 pounds + 40 pounds tank pressure (figure 37) = 43.7 <u>pounds total pressure head.</u>

Total Head

The total head is the sum of (1) the suction distance, (2) the lifting distance (in the case of a deep well pump), (3) the discharge elevation, and (4) the friction head caused by water passing through pipe.

Pump Location

The natural tendency is to locate a pump at least expense with respect to pipe and housing costs. Regardless of where the pump goes, the following points should be kept in mind:

1. The suction line of a shallow well system should be as short and straight as possible.
2. Ample protection should be provided against freezing.
3. The pump should be kept easily accessible for oiling and repairing.
4. Good drainage for excess water should be provided.
5. Allowance should be made for ventilation; excess moisture often causes motor trouble and lengthening of V-belts.
6. Dirt and rubbish should be kept out.
7. The pump should be located high enough that flood waters will not rise above it.
8. The pump should be located so that it would be affected last in case of fire.
9. The pump should be elevated on a rigid base above the floor level.
10. The pressure tank should be raised enough to permit air circulation, which helps to prevent corrosion on the bottom.

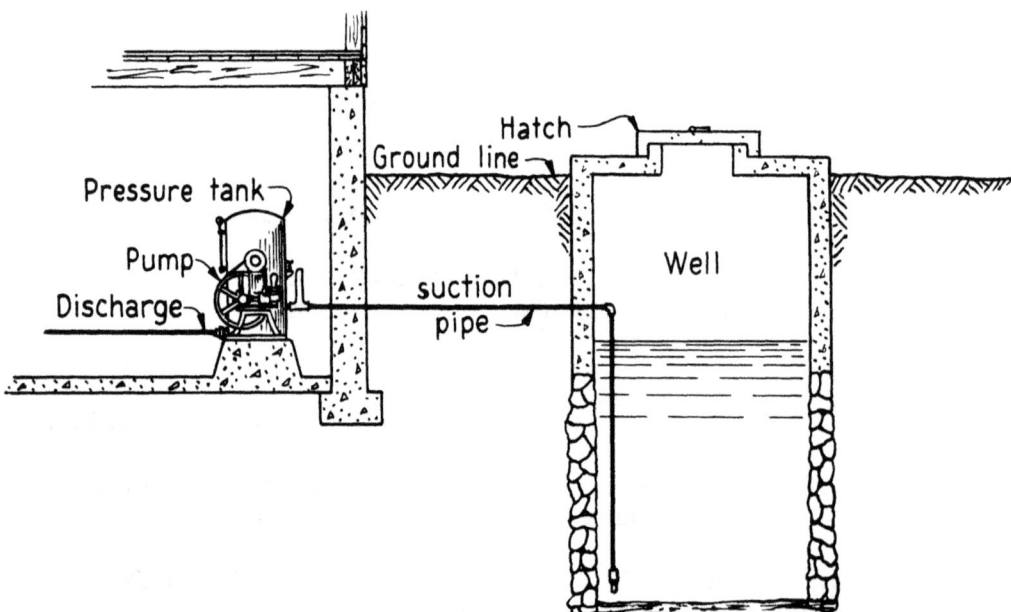

Figure 38. Shallow Well Pump Installed in Basement
(Also Possible with Jet Pump)

Basement Installation

Installation of a shallow well pump or a jet pump in a dry, well ventilated basement with a gravity drain to the ground surface is very satisfactory. A long suction line from well to basement is sometimes a limiting factor; but to some extent this is offset by the fact that the pump is usually three or four feet lower than ground level, if it is set on the basement floor, which gives that much advantage on the suction lift (figure 38). If floor space is limited either type of pump can be mounted on wall brackets.

A level, concrete block 6 to 12 inches high is very desirable for mounting both pump and tank. The height permits easy oil drainage, the pump can be repaired more readily, and the whole assembly is above the level of scrub water and any normal amount of debris that may collect around it

A deep well pump cannot be satisfactorily installed in a basement because the height between floor and ceiling is insufficient for removing drop pipe.

Basement Extension

Figure 39 shows a basement extension. It is simply an extension of the basement beyond the house foundation to form a small pump room. Any type pump may be installed in a basement extension, but it is particularly well adapted to deep well systems.

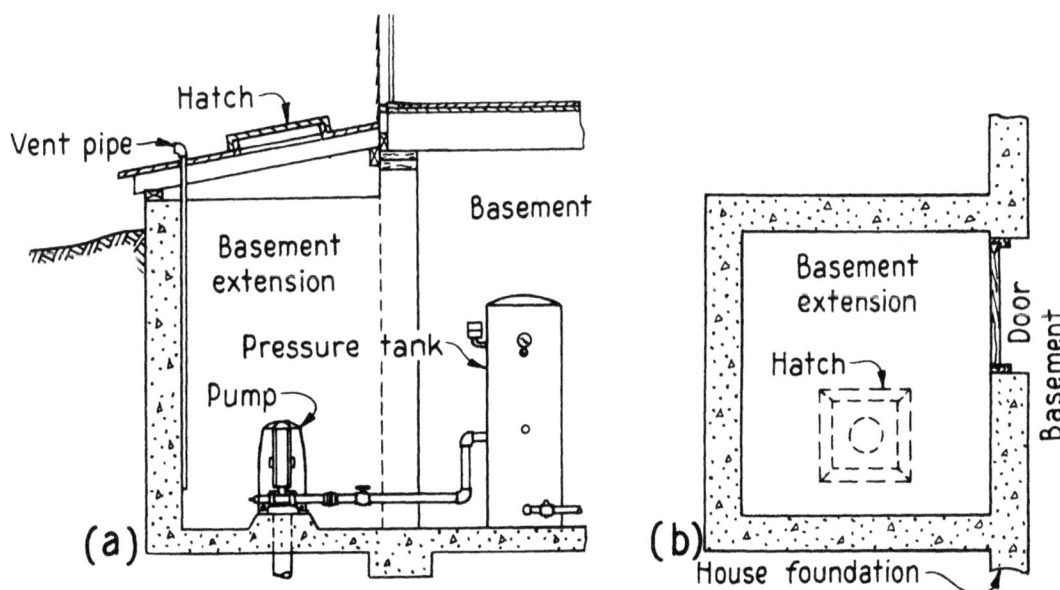

Figure 39.   (a) Basement Extension Type Pump Installation
             (b) Floor Plan

It is usually about 6 feet square by about 5-1/2 to 6 feet deep. The floor should be of concrete and the walls of concrete or other rigid material impervious to water.

The top can be concrete if provision is made for a hatch immediately above the drop pipe. Most installations in the Valley states use a sloping, wood platform cover with a good roof over the platform. The complete platform is set aside when the pump or drop pipe is to be removed.

To aid in lifting the drop pipe many farmers have extended an arm from the frame of the house, to which a rope and pulley may be attached when lifting the drop pipe for repairs.

Adequate ventilation is usually provided from the main basement room unless the basement extension is closed off; in that case a hole several inches square should be inserted in the partition at both top and bottom

Drainage is usually directly into the basement.

The pump should be mounted on a concrete base. The pressure tank may be installed in the basement extension or in the basement proper.

Pit

In some sections the advisability of using a pit is a very debatable sub ject among pump men and health authorities, the principal reason being the abuses in relation to what is considered good pit construction and pump care. In fact, some states are discouraging any type of pit construction. Some of the abuses are:

    1. Little or no drainage. Water wasted from the pump and that accumulating from seepage through the walls often total enough to rise to the level of the well casing, where it enters the well. Besides, some damage is usually done to the pump and the motor.
    2. Improperly constructed pits. Many pits are very poorly built and in some cases no wall or floor is constructed, simply a hole in the ground. Although some installations of this type give no trouble, there are many more that have not been satisfactory.
    3. Insufficient provision to keep out surface water, dirt, and rodents.
    4. Cutting off of well casings even with the floor.
    5. Little or no ventilation.
    6. Neglect of pump and motor.

The principal advantages of a pit are (1) adequate protection against freezing and (2) elimination of a building, which in some cases is desirable from the standpoint of appearances.

Common pit sizes are 5' x 5', 6' x 6', 5' x 7', 7' x 9', by 5-1/2' or 6' deep. However, the dimensions should be figured after the pump and tank sizes are considered. Ample room should be left for easy access during installation and for repair work. The additional room is an advantage, too, if the pump should be replaced by a larger one at a later date.

Most pits are of poured concrete which forms a solid floor and side walls however, they may be constructed with a concrete floor and have brick, stone, tile block, concrete block, or precast concrete walls. Whatever material is used should be watertight. The floor should slope to a drain and the walls should extend a few inches above ground to keep out surface water (figure 40).

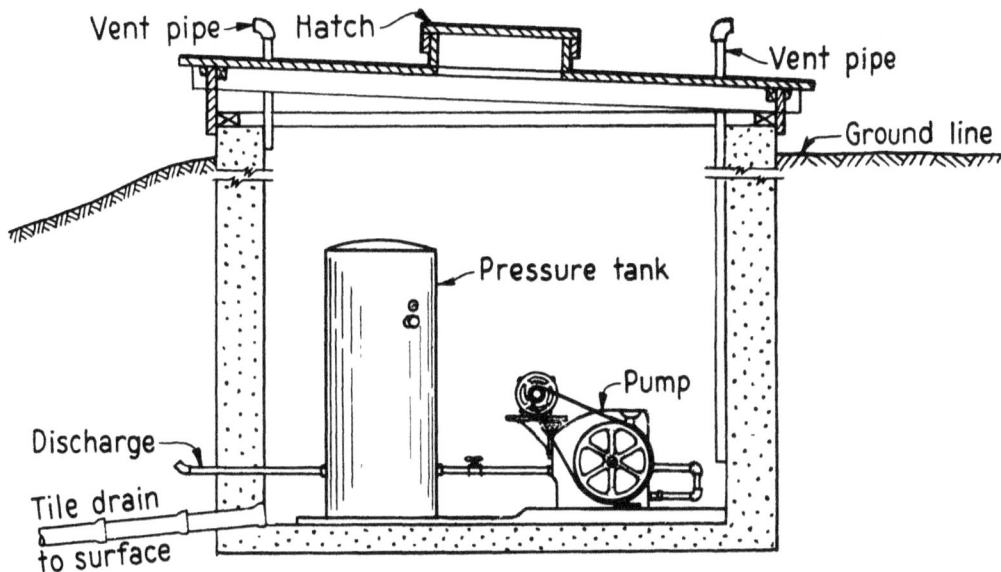

Figure 40. Pit Type Pump Installation

Some pits are drained through a hole in the floor into a sump. This type of sump consists of a hole dug in the ground below floor level, which is filled with gravel or other coarse aggregate that permits the water to escape quickly. However, a sump is limited to use where the ground is loose so that water will soak through it easily, and also where water tables stay below pit levels. In general a sump cannot be recommended.

A tile drain with an open outlet is best. In case that is not possible or practicable, the drain may empty into a drainage line but should never empty into a sewer tile.

Ventilation is at least partially taken care of by the vent system shown in figure 40, which consists of 2-inch or 3-inch pipe or downspout. In very cold weather the vents may have to be closed.

The well casing should extend a few inches above the floor and a concrete pump base should be poured to keep the pump above floor level.

Frostproof Set Length

The frostproof set length is applicable to deep well piston pumps only. As shown in figure 41, the discharge head is lowered into a pit somewhat smaller and more shallow than a pump pit, so that all water connections are below frost level.

With this arrangement the pump head can set above ground in a small shelter house, which makes it popular for gasoline engine use. The arrangement provides easy access to the motor and pumping head, but in other respects it is subject to the same abuses as an ordinary pit. However, if the pit can be arranged as a basement extension, it is considered more satisfactory.

Frostproof set lengths can be installed satisfactorily. It is the fact that they generally are not that has caused them to be viewed with disfavor by health authorities.

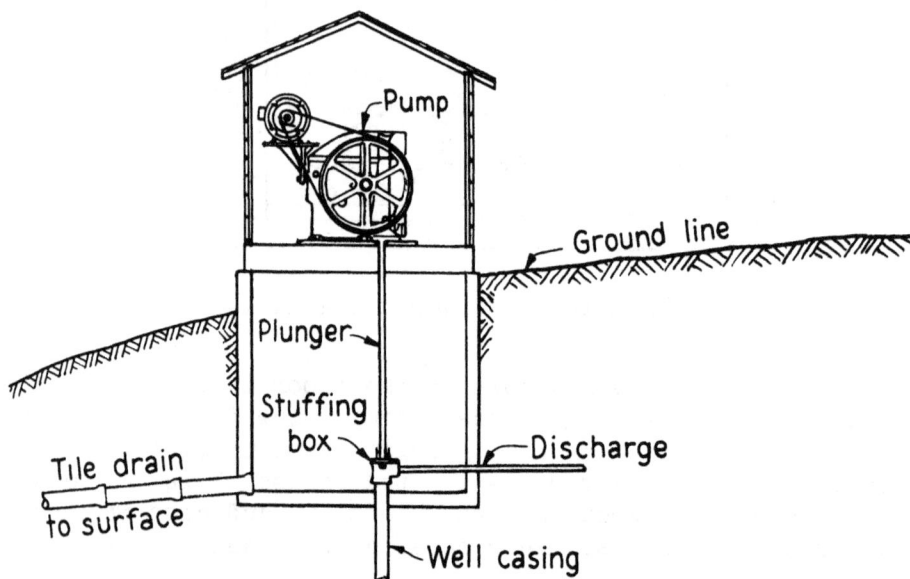

Figure 41. Pump Installed with Frostproof Set Length

Pump House

In many sections pump houses are gaining in popularity through the encouragement of state health departments. A well built pump house costs little if any more than a well built pit, and it offers a number of advantages over a pit. These are:

1. It can be well ventilated.
2. It is easily drained.
3. The pump is readily available for oiling and repairs.
4. The well is easily protected from surface water.

If the pump house is not properly insulated, there is a decided risk of the pump freezing, which may prove costly. However, in the southern states below Tennessee, particularly in the area within 200 or 300 miles of the Gulf of Mexico, insulation is not considered absolutely necessary if the house is tightly constructed. In other areas insulation requirements vary to such an extent that it is difficult to give definite recommendations. Figure 42 gives relative effectiveness of different type walls as heat insulators. Doors and windows are not considered in the graph.

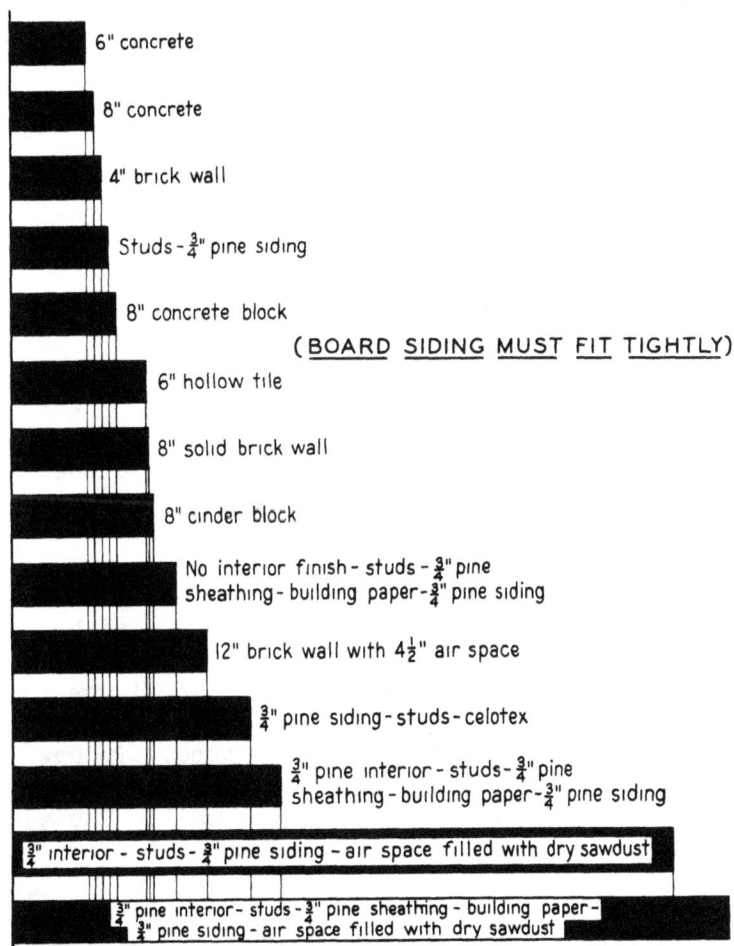

Figure 42. Relative Heat Insulating Effectiveness
of Walls for Pump Houses

From the graph it is quite evident that dry sawdust is an outstanding insulator. However, it has certain disadvantages in that (a) it settles, leaving the upper wall vacant; (b) rodents and termites are readily attracted; and (c) it rots if not protected from moisture.

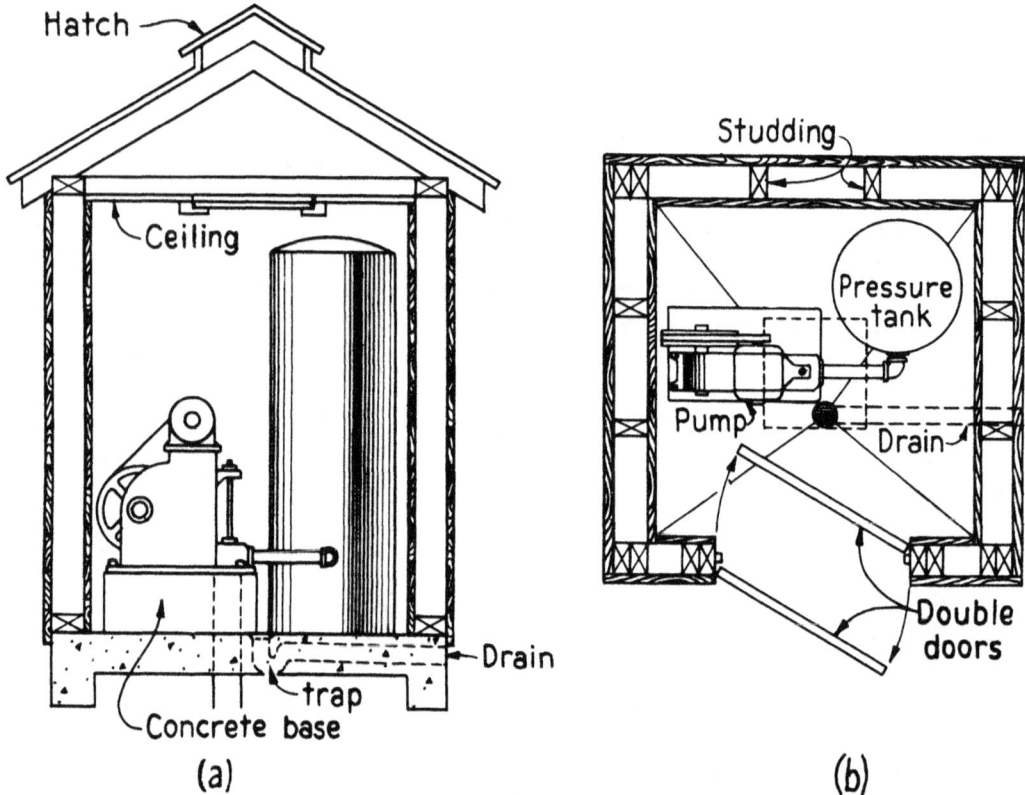

Figure 43. (a) Insulated Type Pump House
(b) Floor Plan

Pressure Tank Location

Where possible the pressure tank should be set close to the pump. A pump installation in a basement or basement extension usually presents no problem, since there is normally ample room for the tank.

It is quite common for the tank to be set with the pump in a pit or insulated well house. This arrangement has the advantage of helping maintain temperatures above freezing since the water in the tank is replaced periodically by comparatively warm well water. However, large tank installations of 80 gallons or over may require a larger or deeper pit or a larger pump house, which means more cost for the structure. To overcome this problem a horizontal tank can be buried in the ground adjacent to a pit or basement, so that one end protrudes through the wall. The air regulator, pressure gage, and pressure switch are attached to the exposed end.

Since the soil around the tank tends to hasten corrosion, even on galvanized tanks, it is well to apply a coating of hot asphalt or a preservative paint before the tank is placed in position.

TABLE V

FARM POWER PUMPS CHARACTERISTIC CHART

Only pumps designed for ordinary farm water supplies are listed below

TYPE PUMP	SPEED	PRACTICAL SUCTION LIFT	PRESSURE HEAD	SUCTION CHARACTERISTICS	DELIVERY CHARACTERISTICS
RECIPROCATING (1) Shallow Well (a) Low Pressure (Type normally used on farms)	Slow	22 to 25 feet	40 to 43 lbs.	Pulsating Positive action within suction limits – little slippage Pumps water containing sand and silt Pumps air for pressure tank	Pulsating (air chamber evens pulsations) Positive delivery, even at high pressure
(b) Medium Pressure	250 to 550 strokes per min.		Up to 100 lbs.		
(c) High Pressure			Up to 350 lbs.		
(2) Deep Well	Slow 30 to 52 strokes per min.	Available for lifts up to 875 feet Suction lift below cylinder 22 feet	Normal 40 lbs. When pressure head must be increased, 2.3 feet is deducted from lift for each pound increase in pressure head	Pulsating Positive action Pumps water containing sand and silt Separate air pump provided for pressure tank use	Pulsating (air chamber and differential cylinder evens flow) Positive delivery
CENTRIFUGAL (1) Shallow Well (a) Straight Centrifugal (Single stage)	High 1750 and 3600 RPM	15 ft. maximum	40 lbs. normal 70 lbs. usually maximum	Nonpulsating Pumps water containing sand and silt Loses prime easily Capacity decreases as lift increases Does not pump air with water for pressure tank. Special equipment required	Continuous, nonpulsating High capacity with low pressure head Capacity reduces rapidly on high pressure head
(b) Turbine Type (Single impeller)	High 1750 RPM	Manufacturers figure 28 ft. maximum at sea level	40 lbs. normal Available up to 100 lbs.	Nonpulsating Capacity reduces as lift increases but not to the extent of a straight centrifugal pump Not satisfactory for water containing sand or silt unless used with settling chamber Pumps air for pressure tank	Continuous, nonpulsating High capacity on low pressure head Capacity reduces on high pressure head but not to the same extent as a single stage centrifugal
JET PUMP Shallow Well and Limited Lift Deep Well	Used with centrifugal-turbine or shallow well reciprocating pump	Maximum around 120 ft. More practical at lifts of 80 ft. or less Creates effective suction 15 to 20 ft. below ejector	40 lbs. normal Available up to 70 lbs. pressure head	Nonpulsating Capacity reduces as lift increases Piston-jet and turbine-jet combinations handle air, other provisions necessary for centrifugal-jet combination	Continuous, nonpulsating High capacity on low pressure head Capacity decreases on high pressure head
ROTARY PUMP Shallow Well	400 to 1725 RPM	22 feet	About 100 lbs.	Positive action Slightly pulsating Increased lift has little effect on capacity Not adapted to water containing sand or silt	Positive, slightly pulsating Increased pressure head has little effect on capacity

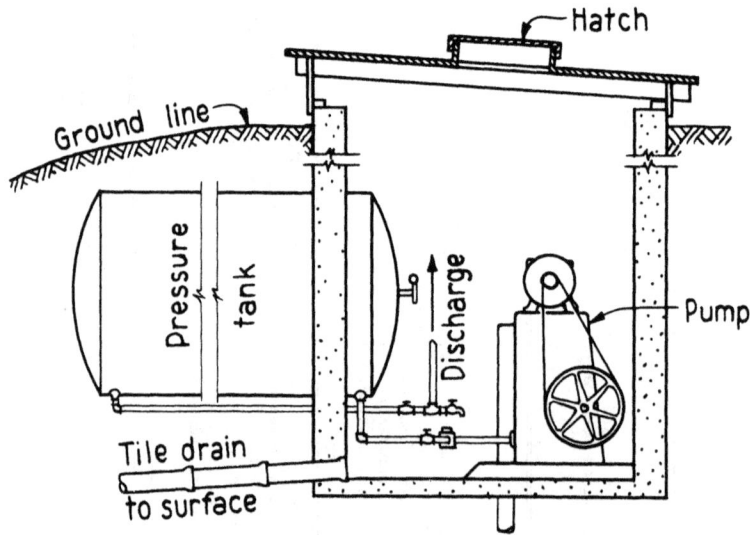

Figure 44. Large Pressure Tanks May Be Installed
Underground to Save Housing Cost.

Wiring to Motor

Pump motors on shallow well systems are often as small as 1/4 hp. or 1/6
hp., and can be successfully operated on 115 volts; however, if the motor
is designed for 230-volt service, it is well to connect it to that voltage. Larger motors should be operated on 230 volts only.* The higher
voltage (1) helps to eliminate motor overheating; (2) permits smaller wire
size, which is a particular advantage when the pump must be located at a
considerable distance from the house; and (3) makes the operation of the
pump motor have little or no effect on the residential lights.

Many farmers have complained that their lights flicker when the pump motor
operates. In most cases the flickering is due to the fact that the motor,
particularly on piston pumps, causes the voltage to fluctuate on the light
circuit.

If the pump is to be used for fire protection, it is best to have it on a
separate circuit. Underground circuits are desirable if the pump is installed in a pit or pump house.

Suction and Delivery Pipe Installation

The suction line of a shallow well system should be installed level (figure 45b) or on a grade draining towards the well. If installed as shown
in 45a, air is trapped in the high section of pipe, causing a limited flow
or no flow at all.

*Note table XVI in the appendix, "Minimum Wire Size Between Power Source
and Motor."

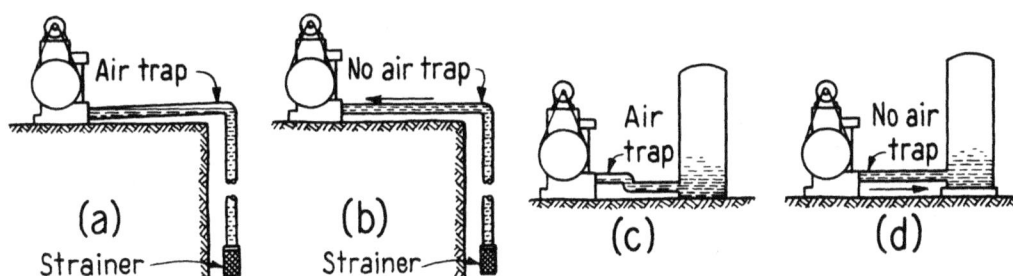

Figure 45.

It is equally important that the discharge line be installed as shown in figure 45d rather than as in c, for the same reasons as discussed for suc tion lines.

When the pipes are installed properly the system can be completely draine if necessary, to prevent freezing.

The following table gives usual depths of underground piping necessary to prevent freezing in various parts of the United States. Since an automat water system is often the only source of water supply, it is well to inst the suction, delivery, and service pipes at least as deep as the minimum depth indicated.

TABLE VI

Usual Depths of Underground Piping for Frost Protection

State	Depth in Feet	State	Depth in Feet
Alabama	1-1/2 to 2	Mississippi	1-1/2 to 2-1/
Arizona	2 to 3	Missouri	3 to 5
Arkansas	1-1/2 to 3	Montana	5 to 7
California	2 to 4	Nebraska	4 to 5-1/
Colorado	3 to 5	New Hampshire	4 to 6
Connecticut	4 to 5	New Jersey	3-1/2 to 4-1/
Florida	1 to 2	New Mexico	2 to 3
Georgia	1-1/2 to 2	New York	4 to 6
Idaho	4 to 6	North Carolina	2 to 3
Illinois	3-1/2 to 6	North Dakota	5 to 9
Indiana	3-1/2 to 5-1/2	Ohio	3-1/2 to 5-1/
Iowa	5 to 6	Pennsylvania	3-1/2 to 5-1/
Kansas	2-1/2 to 4-1/2	Tennessee	2 to 3
Kentucky	2 to 3-1/2	Texas	1-1/2 to 3
Louisiana	1-1/2 to 2	Virginia	2 to 3-1/
Maine	4-1/2 to 6	Wisconsin	5 to 7
Massachusetts	4 to 6	Wyoming	5 to 6
Michigan	4 to 7	Dist. of Columbia	- -- 4
Minnesota	5 to 9		

Sanitary Well Caps

Too often the top of a well casing is left open, permitting the entrance
of dirt, rodents, insects, or surface water. Deep well pumps are often
provided with a collar, which is part of the pump casting and will fit
over the casing enough to provide some protection. Various other arrangements can be used, but one of the most sanitary is a cap of the type shown
in figure 46.

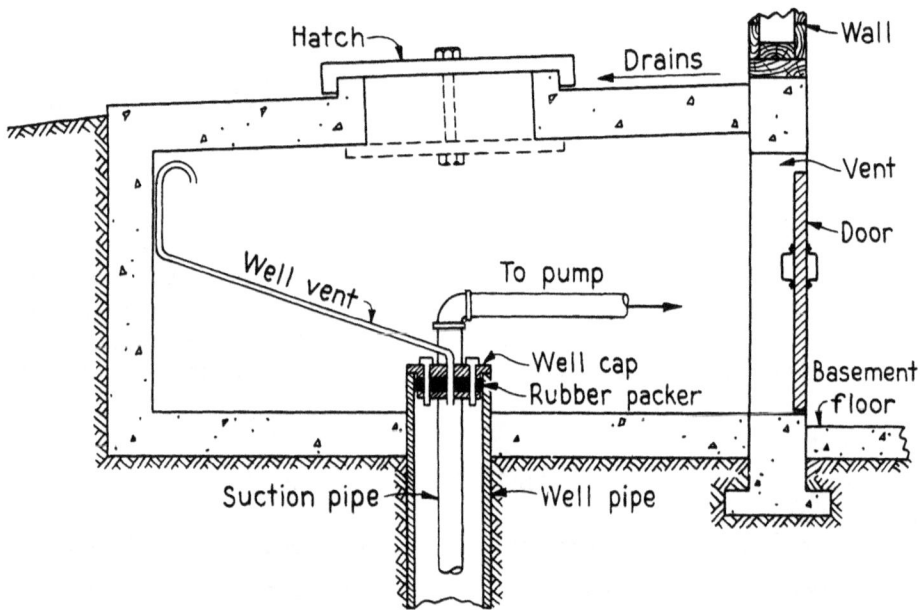

Figure 46. A Tight Well Seal Is Excellent Insurance
Against Contamination. Available for Use
with Deep or Shallow Well Pumps.

The rubber cushion is squeezed between two cast iron plates and forced
tightly against the well casing and drop pipe. The resulting seal is
watertight.

If the water in the well tends to lower more than 10 feet during a pumping period, the well should be vented, as shown in figure 46, to permit

If the water in the well tends to lower more than 10 feet during a pumping period, the well should be vented, as shown in figure 46, to permit
entrance of air.

CHAPTER VI

HYDRAULIC RAMS

The hydraulic ram has a wide application in many sections of the country, and where it can be used satisfactorily it is quite practical. Although simple in operation it is not well understood by farm people, which may account for its comparatively limited acceptance in many sections where it can be used.

Ram Operation

Single Acting Operation

Most of us have probably noticed when we suddenly close a faucet supplying water under pressure that there is a thud or hammer, usually accompanied by pipe vibrations. The hammer caused in that manner is the principle underlying hydraulic ram operation.

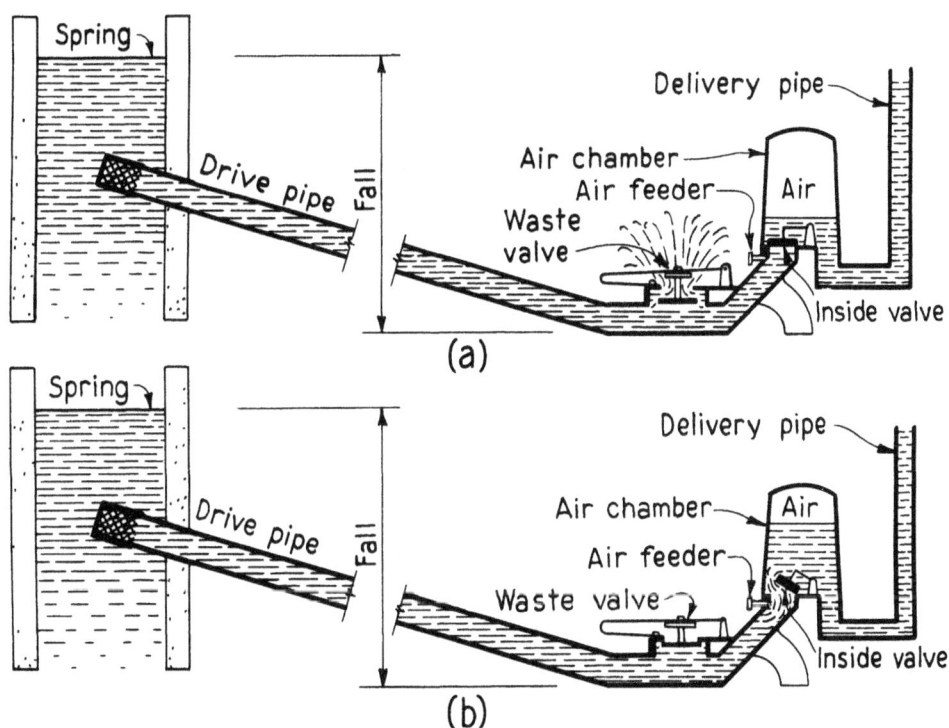

Figure 47. Ram Operation
(a) Water Rushes Down the Drive Pipe and Out the Waste Valve.
(b) Waste Valve Closes and Water Forces into Air Chamber.

To illustrate, in figure 47a the water passing from the spring down the drive pipe to the waste valve quickly increases in velocity to the point where the waste valve suddenly closes, because of the friction of the water passing it and the increased pressure.

A moving water column, like any other moving body, builds up momentum so that a sudden stoppage means a tremendous thrust on the discharge end, which varies with the length and size of the column. The thrust is partly due to the head of water, but it is largely due to momentum. The same instant (figure 47b) the waste valve closes on the ram the check valve is forced open and water enters the air chamber, where the air is compressed until it is equal to the thrust pressure from the drive pipe, when the check valve closes. When this happens the pressure in the drive pipe is still greater than the head pressure alone. Consequently, there is a rebound or backward flow up the drive pipe, which not only relieves the excess pressure but tends to create a vacuum, causing the waste valve to open by its own weight, thus permitting the water to reestablish a flow back down the drive pipe.

The rebound performs a second function, in that it offers an opportunity to introduce a small quantity of air into the air chamber. This is accomplished by locating a small snifting valve, or air feeder, in the base of the ram below the air chamber. On the rebound a small quantity of air is forced through the valve opening and on the next thrust is forced into the air chamber along with the water.

The action of the ram is rather rapid, the complete cycle of operation being completed at the rate of 25 to 100 times per minute, depending on the waste valve adjustment and characteristics of the installation.

Hydraulic Ram Adaptability

Minimum conditions for ram operation are considered to be:

1. Not less than 18 to 20 inches fall (24 inches or more is preferable).
2. Twenty-five feet of drive pipe.
3. Flow of at least 1-1/2 gallons per minute.

Of course under such minimum conditions of operation the amount of water delivered and the height to which it is delivered are proportionately limited. For example, under the minimum conditions just mentioned, if water is lifted 10 feet an efficient ram would probably not deliver more than 9 gallons per hour. If the lift is lowered, the amount of water delivered increases. With increased lift the opposite is true. If the flow can be increased, the drive pipe lengthened, or the fall increased, a greater delivery of water can be expected.

There are also maximum conditions under which commercial rams operate, which are normally considered to be:

1. 30-foot fall. (Some manufacturers prefer that their rams operate on a maximum of 16 feet fall.)
2. 250 feet of drive pipe.

To some the maximum limitations may seem rather low, but experience has proved that when these conditions are exceeded the momentum of the water is so great that an ordinary ram may be damaged by the resulting thrusts. In fact, with installations where the fall is greater than 15 to 16 feet, some farmers have experienced difficulty in making the ram operate. The weight and velocity of the water becomes so great that the waste valve is held closed.

Water Source. Hillside springs are the most common source of supply. Artesian wells can be used if there is sufficient flow and if the water rises sufficiently high in a standpipe or tank to give the necessary head. When there is a limited supply of good spring water and an adequate supply of creek water, arrangements can often be made to pipe both to a lower level and make the creek water pump the pure water by means of a double acting ram.

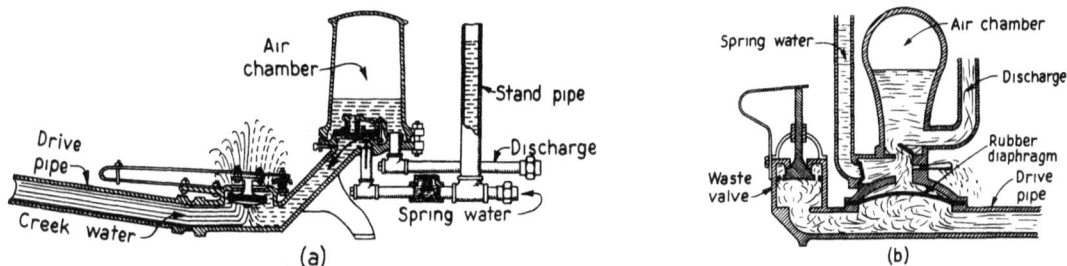

Figure 48. (a) Double Acting Ram Which Uses Creek Water to Pump Spring Water
(b) Double Acting Ram with Diaphragm Separating Creek from Spring Water

As shown in figure 48a, the spring water actually enters the base section of the ram. The action occurs during the rebound previously discussed. If the ram is working properly, that portion of the spring water forced into the air chamber has had no contact with the creek water. It is necessary to waste about one-third of the spring water through the waste valve in order to accomplish this action. However, if this type ram is not working properly, the spring water and creek water will mix. This has caused some question on the part of state health departments as to the advisability of installing it. At least one manufacturer has attempted to overcome this difficulty by using a rubber diaphragm to separate the spring water from the creek water but still impart pumping action to the spring water from the drive pipe (figure 48b).

Pressure at the Faucet. The same three systems are used as were mentioned in the water system discussion. If there is an abundance of water, no storage tank is necessary. The water can be piped to the house, then to various buildings on the farmstead, and the remainder discharged into a ditch or tile drain.

A gravity tank is commonly used where the supply is not sufficient to take care of peak demands.

A pressure tank can be used if the ram will develop sufficient discharge head without too much sacrifice of the volume pumped. A relief valve of sufficient capacity to provide quick release of excess pressure should be installed on the delivery side. To determine the total discharge head when a pressure tank is used, follow the same procedure as with pumps, using table IV. It is well to install a pressure tank large enough to hold one full day's supply.

Installation of Ram

Necessary Data

Before a hydraulic ram is installed certain necessary data should be secured, as indicated in figure 49.

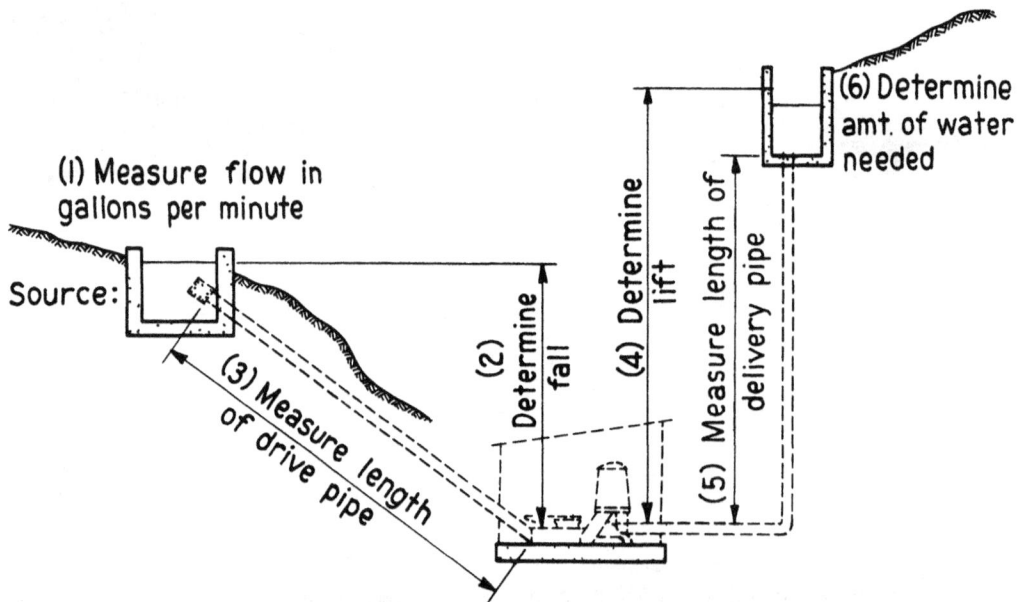

Figure 49. Necessary Data to Secure Before Ordering
a Ram

Care should be taken to figure for a minimum flow in case the flow fluctuates during the year.

TABLE VII

Typical Cases Showing How Much Water Rams
Will Pump under Varying Conditions

Supply Water Gals. per Min.	Feet of Fall to Ram	Vertical Elevation (Feet)	Gallons Delivered per Hour	Proper Length Drive Pipe (Feet)
5	3	40	15	40
10	4	80	20	80
15	20	300	40	250
30	15	100	180	100
45	35	200	315	200
80	50	500	300	250
150	20	100	1200	100
250	15	60	2500	60

A formula for figuring approximate capacities of hydraulic rams under given conditions can be found in the appendix.

The proper size ram to use will be determined by the manufacturer. Pointers on the actual installation of the ram are:

Drive Pipe

1. Pipe should normally be 10 to 20 per cent longer than vertical height of delivery line.
2. Pipe must be straight. Avoid any humps, dips, or bends.
3. Upper end must be submerged at least one foot.
4. Strainer on upper end should have holes with a total area of 3 to 5 times cross-sectional area of drive pipe.
5. Strainer should be 6 to 8 inches above bottom of reservoir.
6. Full-way gate valve is desirable at lower end of drive pipe. (Install with stem horizontal to floor.)
7. If it is necessary to pipe the water some distance to obtain the necessary fall, a supply pipe and standpipe may be used, as indicated (figure 50).
8. A slope of 2-1/2 inches per foot of length is considered most efficient.
9. The length of drive pipe should not normally be less than 5 nor more than 10 times the fall from source to ram.

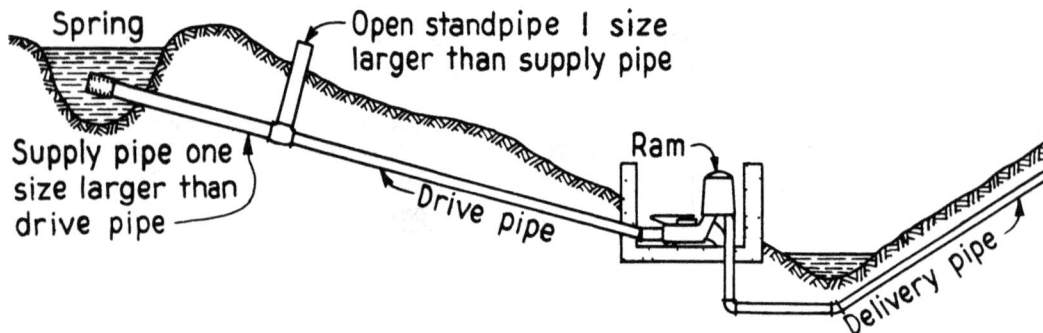

Figure 50. If Distance from Spring to Ram Exceeds 250 Feet, a Standpipe Is Used.

Ram

1. Mount ram solidly on concrete slab or stone.
2. Allow at least 18 inches clearance on all sides of ram.
3. 4-inch drain tile is commonly used to drain ram pit (grade 1-1/2 to 2 inches per 100 feet).
4. Build pit walls so that they extend above flood level. If waste outlet is not too large under flooded conditions, the ram will continue to operate and the waste water level will be higher inside the pit than the flooded water level outside.
5. Complete side walls and roof after ram is installed and pipe connections made.
6. Slope floor towards drain.
7. Pit is usually made of concrete.

Delivery Pipe

1. Keep delivery pipe as short and straight as possible.
2. Maintain pipe on an upward grade; avoid high or low points.
3. Do not use pipe smaller than discharge tapping on ram.
4. Use pipe compound on threaded joints.
5. Do not cover pipe until ram is started so leaks can be detected.
6. Full-way gate valve at ram is convenient. (Install with stem horizontal to floor.)
7. Cover sufficiently to protect against freezing.

P A R T II

CHAPTER VII

PLUMBING--WATER HEATERS--WATER TREATMENT

PLUMBING

Pipe and Pipe Fittings

Wrought Iron and Steel Pipe

Both wrought iron and steel pipe are used in farm plumbing but few farmers are able to tell the difference by appearances. Wrought iron is more expensive and consequently less popular. Both are available in a black or galvanized finish, the latter being the most durable. The galvanized finish is a zinc coating applied to the pipe; it acts as a protective covering and helps prevent corrosion.

Copper bearing steel pipe is ordinary galvanized steel pipe containing about .2 per cent copper by weight, which adds to its capacity to resist corrosion.

First quality wrought iron and steel can be recognized by the manufacturer's marking on each pipe, which indicates it is of high quality, is uniform in size and weight, and meets Standard Specifications as established by the American Society for Testing Materials.

For underground use the life of galvanized wrought iron and steel pipe is generally considered to be from 15 to 20 years. Black pipe should not be used on water lines either above or below ground.

Malleable iron fittings are used with both wrought iron and steel pipe. It differs in structure from both types of pipe material in that it is tougher and can be hammered or bent to a limited extent. Cast iron fittings are also widely used.

Couplings and nipples are usually of steel.

Copper Pipe or Tubing

Copper pipe has gained wide acceptance during the last few years for farm plumbing. It is available in three different weights or wall thicknesses, and with two of the three types it can be secured in either hard 20-foot pipes or soft 30- to 60-foot coils.

Type K, hard or soft, is heavy gage, used for underground piping, high pressures, or pump suction lines.

Type L, hard or soft, is medium gage, commonly used for house supply lines

Type M, hard only, is for house supply lines but not generally recommended.

Copper tubing, although somewhat more expensive than wrought iron or steel pipe, has certain advantages:

1. It is not as subject to corrosion as iron or steel piping.
2. Connections are easily made with solder or flange type joints.
3. Pipe a size smaller can be used for the same water delivery.
4. The soft tubing can be bent to miss obstructions and thus eliminate fittings.
5. The soft tubing will stand from 4 to 10 freezes without bursting.

For underground use the type K tubing is excellent. Some plumbers estimate it will last as long as 100 years.

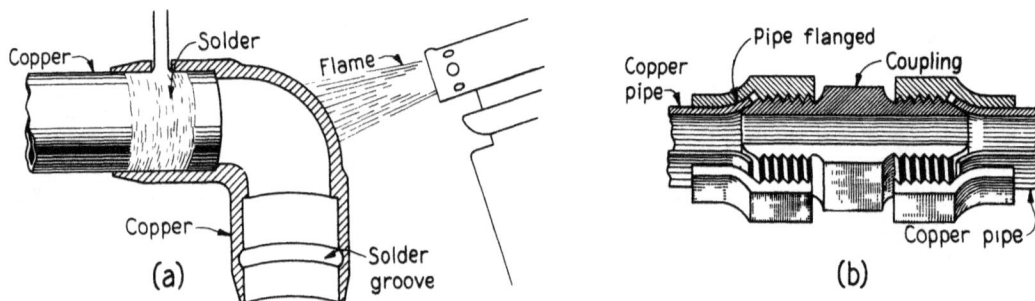

Figure 51. Copper Pipe Joints
(a) Solder or Sweat Type
(b) Flange Type

First quality copper tubing can be recognized by the manufacturer's trademark or brand stamped on the tube. The stamp indicates the tubing was manufactured according to specifications established by the American Society for Testing Materials.

Special fittings are available for connecting copper piping to iron piping

Brass Pipe

Brass pipe is not generally used in farm plumbing because of cost. It can be threaded to make connections. It is corrosion resisting, smooth, and neat appearing. Since it can be obtained in smooth, polished, chromium or nickel plated finishes, it is sometimes desirable for exposed work.

Brass fittings are available in practically all the malleable iron types, and should be used with brass pipe.

Cast Iron Pipe

Cast iron pipe is used for waste and venting purposes and hence is usually used in larger sizes than other types of pipe. It comes in 5-foot lengths and in three weights--extra heavy, medium, and standard. The latter is most commonly used. The pipes usually come with a coating of asphaltum or coal tar-pitch.

Various types of fittings are available for connections to closets, sinks, lavatories, etc. Instead of standard threaded connections, the joints are made watertight by inserting the small end of one pipe into the bell of the second, packing oakum in the joint and pouring hot lead into the bell. The resulting joint is airtight and watertight (figure 52).

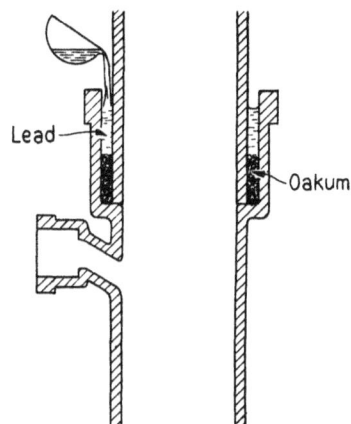

Figure 52. Cast Iron Soil Pipe. Molten Lead Poured Over Tamped Oakum Forms the Seal.

Lead Pipe

Lead pipe is available and used to a very limited extent, mostly where the water is corrosive on other types of pipe.

Pipe Insulation

Heat Insulation

Heat insulation is usually secured by wrapping about 3/4-inch asbestos corrugated covering on hot water pipes inside buildings. The same covering will work on overhead pipes running outside from one building to another if it is protected from weather conditions by a metal or wood cover. However, for most farm installations hot water service between buildings may be so intermittent that there is some risk of the pipe line freezing in cold weather.

Figure 53. Asbestos Insulation for Hot Water Pipes

Insulation to Prevent Freezing

Most farm homes have no protection for cold water pipes within buildings, and often none is needed if the foundation and walls around the pipes are tight or if a furnace is installed. If protection is needed, it is common practice to use 1-1/4-inch felt covering on the pipes.

Some farmers have used soil heating cable by wrapping it around the pipe and connecting the ends to an electric circuit. The cable consists of a small heater wire in a lead sheath. A 60-foot length supplies about 400 watts of heat, which is sufficient to prevent pipe freezing in most cases unless the pipe is exposed to air currents.

Underground cold water pipes are adequately protected if they are laid well below freezing level, as indicated in table VI.

A suction line or delivery pipe which must pass through an exposed space between ground level and the floor of a building constitutes a real problem which many farmers have not solved satisfactorily. One of the more successful systems is shown in figure 54.

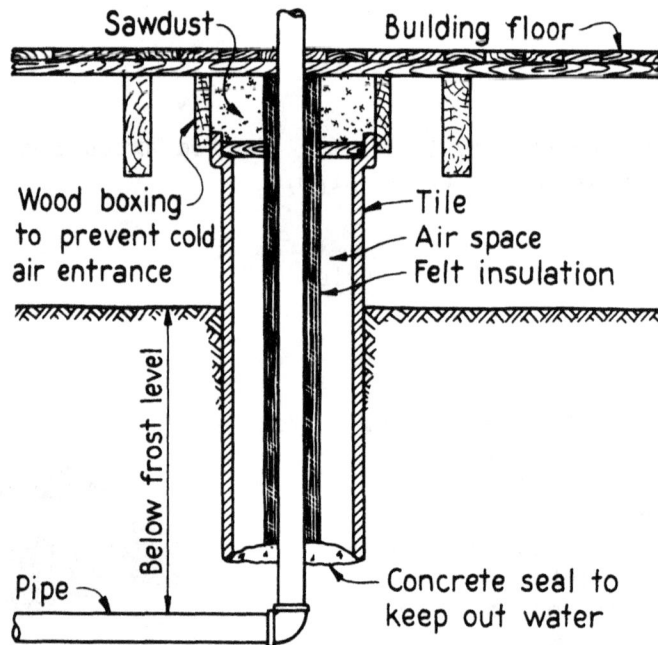

Figure 54. A Durable Protection for a Cold Water
Riser between Ground and Floor

Sawdust is used with varying degrees of success. If it can be kept dry, and additional sawdust added each year as it settles, it is quite effective; however, it provides a refuge for termites.

Soil piled to a 1-1/2- or 2-foot thickness around the pipe is often effective, but it too will settle, leaving the upper part of the pipe
exposed.

It is sometimes desirable to run a hot water line underground between
two buildings in order to prevent pipe freezing when hot water is not
being used.

For example, if there is a source of hot water in the farm house and hot
water is needed in a milk house nearby, it is possible to supply water
through an underground pipe. However, if the pipe is laid directly in
the ground, the damp earth will take up an excessive amount of heat. By
supporting the pipe on wood blocks in 4-inch tile, as shown in figure 55,
and draining the tile system, it is possible to have a satisfactory installation.

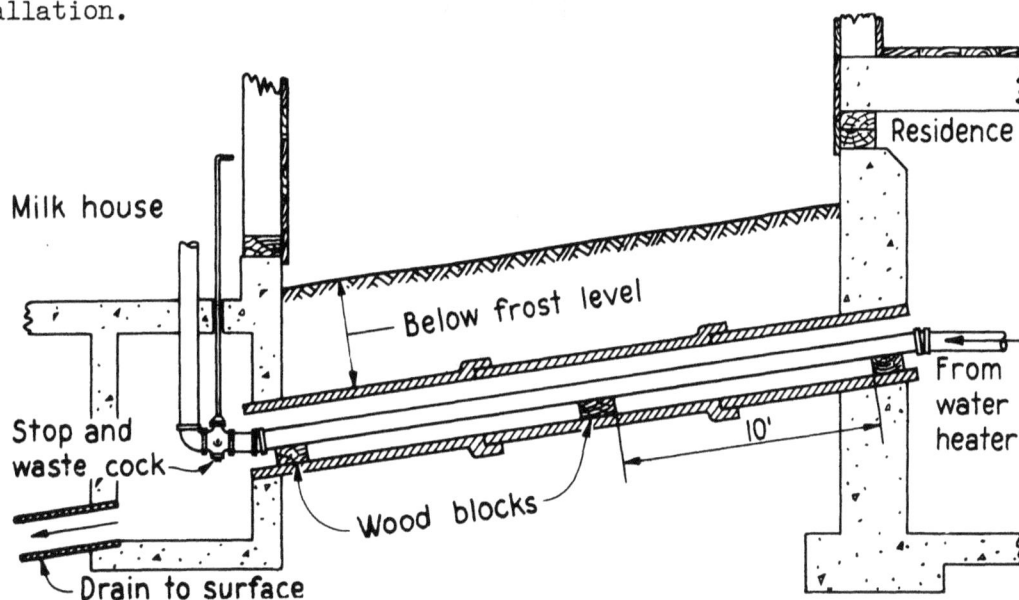

Figure 55. Underground Hot Water Line Extension for
Intermittent Hot Water Service

Supply Piping

Supply piping is that portion of the system that supplies water under
pressure to the various outlets and fixtures about the home and farm.
Three-quarter-inch galvanized steel pipe is commonly used for household
supply lines. On very short runs and sometimes on hot water lines 1/2-
inch pipe may be used. Copper supply lines can be a size smaller than
steel pipe.

If a supply line is to be long, as is often the case with piping between
the house and the barn, the size of pipe should be determined by use of
the friction tables in tables IV and VIII. Too often pipe sizes are guesse
at and a small pipe is installed, resulting in inadequate water delivery.
In fact, it is not uncommon to find long supply lines of small pipe that
will deliver only a drip at the far end.

70

<u>Air Lock</u>

When following surface contour in laying a long supply line underground, the pipe sometimes has one or more high points where air will collect. The air will gradually restrict the flow and may accumulate sufficiently to stop all water passage. The condition can exist with spring supply lines as well as those connected with pressure tanks or hydraulic rams (figure 56).

Figure 56. Air Lock, Common Where Pipe Is Not Laid
on a Grade

To overcome the condition a faucet or air cock can be installed on the high points and the air released occasionally.

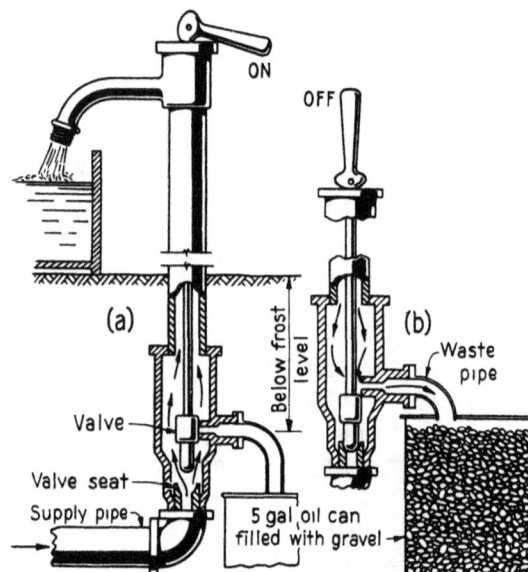

Figure 57. Frostproof Yard Hydrant for Outside
Watering Purposes
(a) "On" Position with Waste Closed
(b) "Off" Position with Waste Open

Waste and Vent Piping

A considerable portion of the cost in making a farm home modern is in the waste and vent piping. Since the operation of the system is not generally understood, it is sometimes difficult for farm people to justify in their own minds the cost necessary for a good plumbing system. Since there is no plumbing inspection in rural sections, farm people are handicapped by not knowing when the waste and vent piping are installed properly. We shall attempt to give here a few of the principles of operation.

Siphon

It is necessary to know first how a siphon works, since the sanitary system is designed either to create or to prevent a siphon condition, as will be studied later.

Most of us have siphoned water from a jar or tank through a rubber hose, and have learned that two conditions are necessary before water will siphon: (1) all air must be exhausted from the hose; and (2) the discharge end must be lower than the water level in the tank from which water is removed.

When such a condition exists the water in leg B (figure 58) is heavier than the water in leg A, resulting in a discharge of water through leg B. The discharge tends to create a vacuum at C, which permits atmospheric pressure to force water from the container up leg A and over into B to replace the discharged water. Thus the operation is continuous as long as the condition exists.

When the water level lowers until leg A is as long as B, the flow will stop

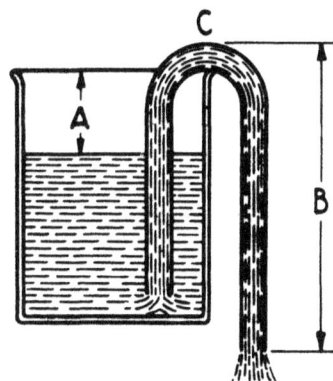

Figure 58. Siphon Action

Stack, Trap, Venting

The stack is the vertical portion of the sanitary system into which waste is discharged and the upper part of which vents foul odors and sewer gas above the roof of a building.

A trap is a fitting or device, usually near a fixture, designed to retain a small amount of water which acts as a seal to prevent the passage of sewer gas into the house.

A vent pipe is a ventilating pipe which prevents water in a trap from being siphoned and which also prevents a back pressure of sewer gas, as will be studied later.

72

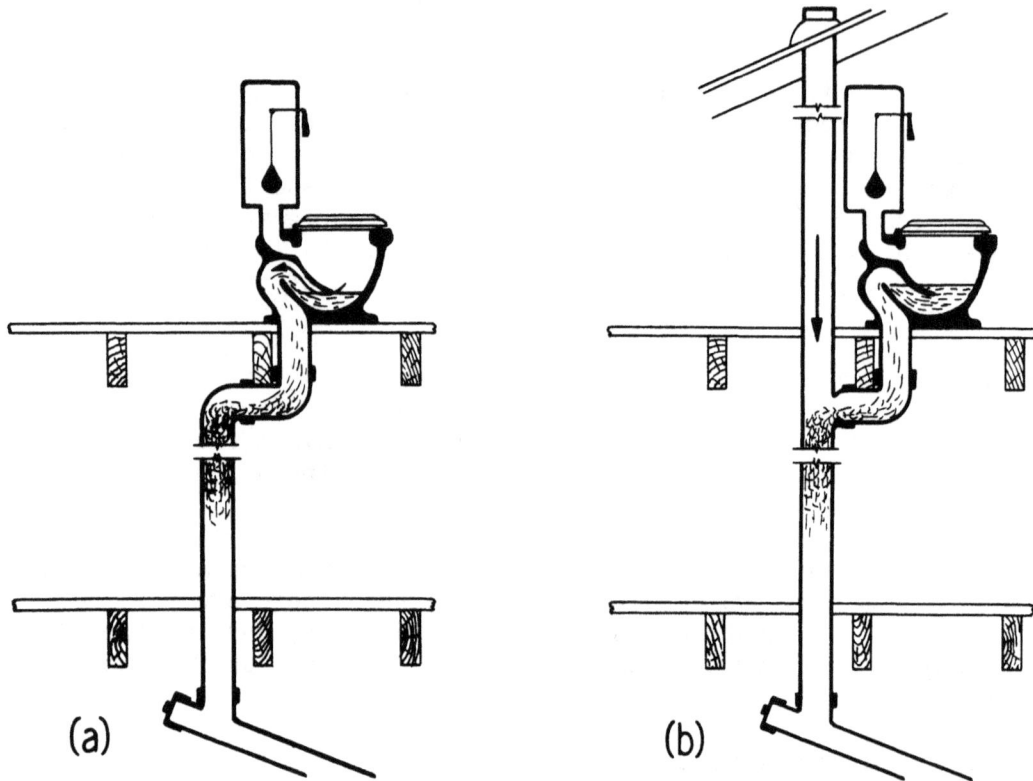

Figure 59. (a) Unvented Stack
            (b) Properly Vented Stack

To understand more clearly the operation of a sanitary system, note figure 59a. The closet is connected directly to the vertical stack. When the closet is flushed, the air ahead of the discharged water is compressed, while the space back of the discharge has less than atmospheric pressure, causing air to rush through the trap. The sudden rush of air through the trap causes a loud sucking noise and tends to carry enough water from the trap to destroy the seal. If the water seal is destroyed sewer gas will pass through the trap into the room.

In order to overcome such a condition the stack is vented through the house roof (figure 59b). With this kind of arrangement the tendency to create a vacuum back of the discharge is overcome by air entering the stack, as indicated.

From the discussion above it might be assumed that all problems related to venting were taken care of by extending the stack through the house roof. However, the condition shown in figure 60a presents still another problem. Note that the position of the closet discharge is still the same, but that there is a second entrance from a lavatory on the first floor of the house. As shown in the illustration, when the discharge from the closet passes the end of the lavatory discharge a suction is

created in the lavatory drain, tending to destroy the seal in the lavatory trap. Air passes through the trap, causing a sucking noise and a
portion of the trap water is drawn into the stack.

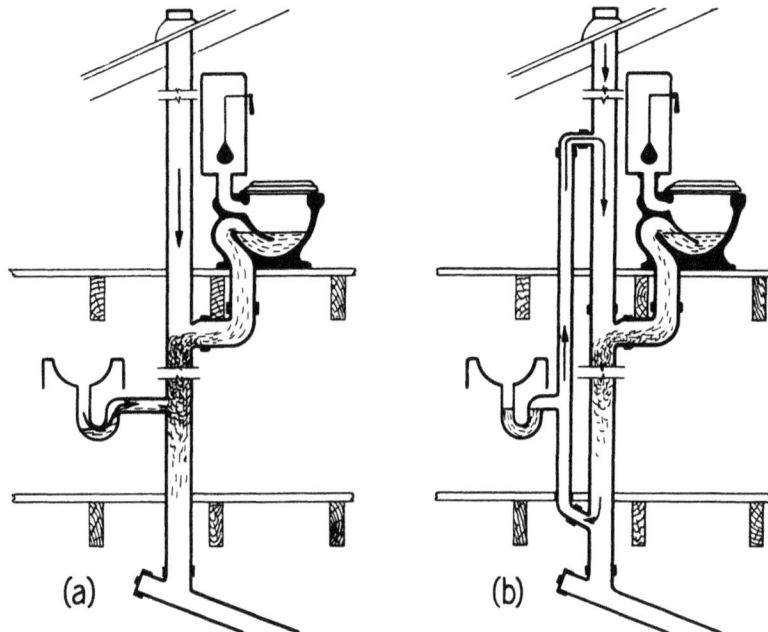

Figure 60. (a) Lavatory Not Properly Vented
            (b) Lavatory Properly Vented by Reventing

The condition in figure 60a can be
readily overcome by reventing, as
shown in figure 60b. This last
step may not be necessary in some
homes, but there is a tendency to
omit this feature when it is needed if the cost begins to increase.

An example of an installation where
reventing is not essential is shown
in figure 61.

A more elaborate plumbing system,
illustrating reventing and two
stacks, is shown in figure 62.

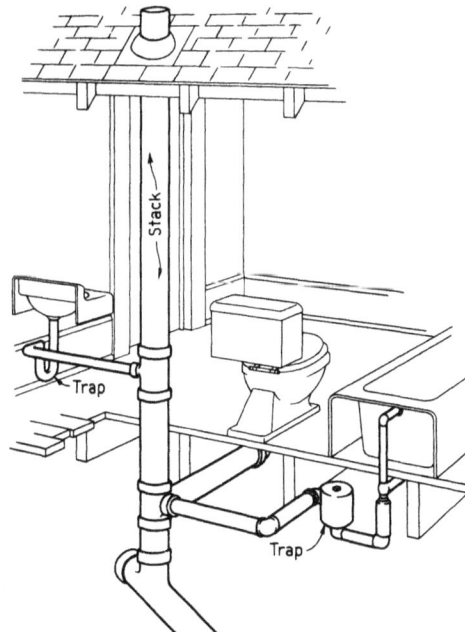

Figure 61. An Arrangement where Reventing Is
            Unnecessary as Long as the Fixtures
            Are within 5 Feet of the Stack.

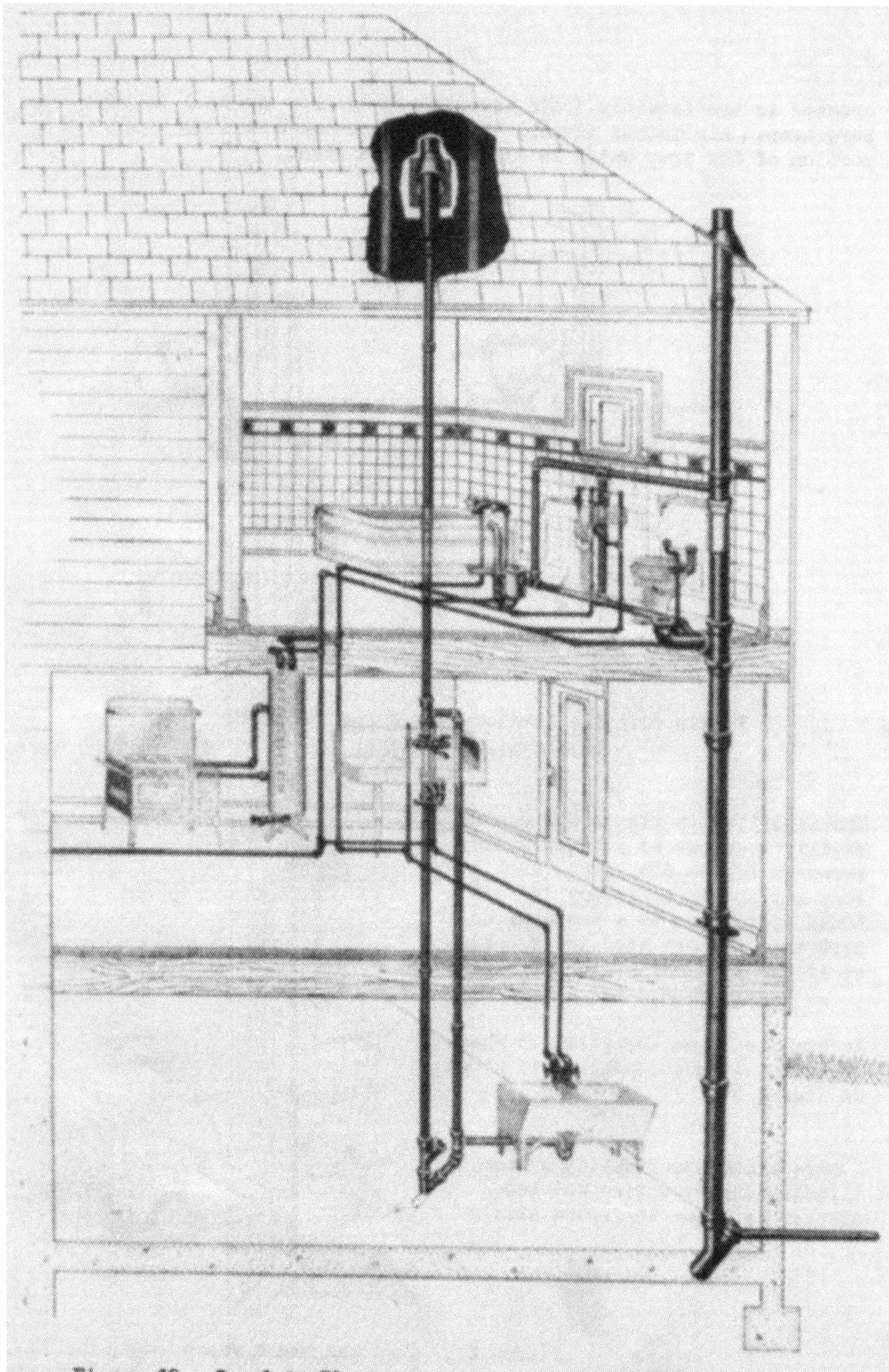

Figure 62. Complete Plumbing System with Revented Waste Lines

Plumbing Precautions

Probably the best insurance of a good plumbing job is to employ a plumber who is well informed and experienced in the trade. There are certain observations that can be made, however, which give some assurance of good workmanship.

1. The supply pipes should slope so that they may be completely drained from a low point.
2. There must be a good watertight connection between the fixture outlet and the drain pipe.
3. The venting should be checked in order to avoid such conditions as indicated in figure 60a.
4. The soil pipe should be checked to make certain it is gastight and watertight.
   a. Oakum should be well tamped into each joint.*
   b. Lead, not cement, should be used to seal the joint.
   c. The whole system can be tested by plugging all outlets and filling the stack and vent pipes with water, entering the water at the top of the stack. The water level should not recede.
5. Care must be used in making threaded joints:
   a. When pipe is cut with pipe cutters there is a burr on the inner edge that should be reamed out (figure 63).

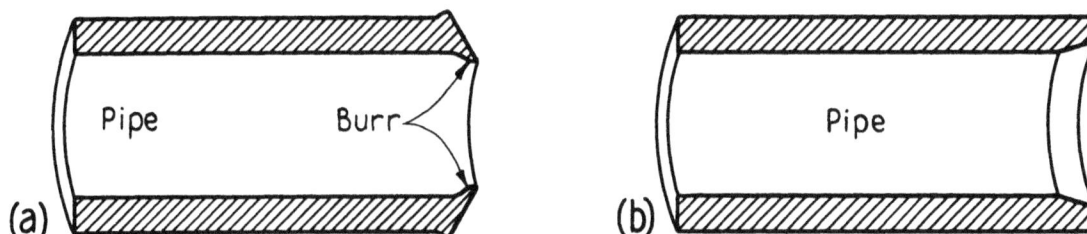

Figure 63. When Pipe Is Cut the Burr Should Be Removed.

   b. Threads can be cut too short or too long. Correct length is shown in figure 64a, and a magnified view of the threads in figure 64b.

*A waste system that does not use oakum and lead joints, known as the Durham waste and vent system, is also used to some extent in homes. It consists of galvanized steel pipe and cast iron drainage fittings with threaded joints which permit tightening with a wrench. It is well adapted for those who may feel skilled enough to install their own systems. The central component parts are shipped assembled from the factory and the balance is "tailored" to fit the building.

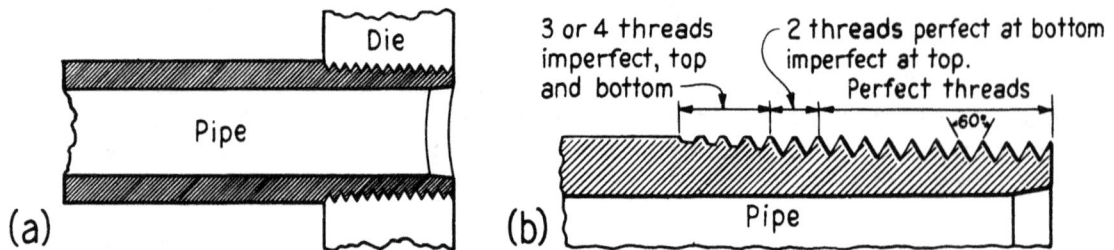

Figure 64.   Correct Pipe Threading

6. Pipe compound should be used on all threaded joints and
    applied to outside threads only.

WATER HEATERS

One of the most neglected features of farm plumbing is provision for adequate hot water service. Although there are a great many ways water can
be heated, there are relatively few means available to farmers that are
inexpensive and automatic at the same time. This fact probably accounts
for intermittent hot water service in most farm homes.

Types of Heaters

Range Boiler

Most homes at present use a range boiler in connection with one or several heat sources. The boiler itself is simply a 30-gallon hot water
storage and seldom is the heat applied directly to it except with certain
gas heaters. Most common sources of heat are from a water front in the
kitchen stove, a furnace coil, a laundry stove with a water jacket, or a
small boiler; side arm heaters are used in connection with artificial or
natural gas, kerosene, gasoline, and, less frequently, electricity.

Whatever type is used the water circulates from the bottom of the range
boiler through the coil and back to the top of the boiler (figure 65).
The fact that warm water is lighter in weight than cold causes it to rise
as it is heated in the coil; then cold water replaces it. The heated
water tends to remain at the top of the tank; consequently the hot water
outlet is at the top.

The fact that the incoming cold water enters the top of the tank seems
inconsistent even though it discharges towards the bottom. It is evident
that the cold water pipe, passing through the hot water layer, tends to

cool the heated water. There is also a small quantity of cold water discharged into the hot through a small hole called a siphon breaker. The
arrangement is a safety precaution to avoid siphoning the water out of
the tank in case a valve lower than tank level is left open at a time
when the water supply line is disconnected and air is permitted to enter.
Under such circumstances, without the siphon breaker the water would be
siphoned from the tank. If heat was being applied to the coils and cold
water was reentered through them, the resulting steam would probably
cause an explosion.

If cold water entered the bottom of the tank directly, there is still a
possibility that the tank might be drained if the supply line was disconnected.

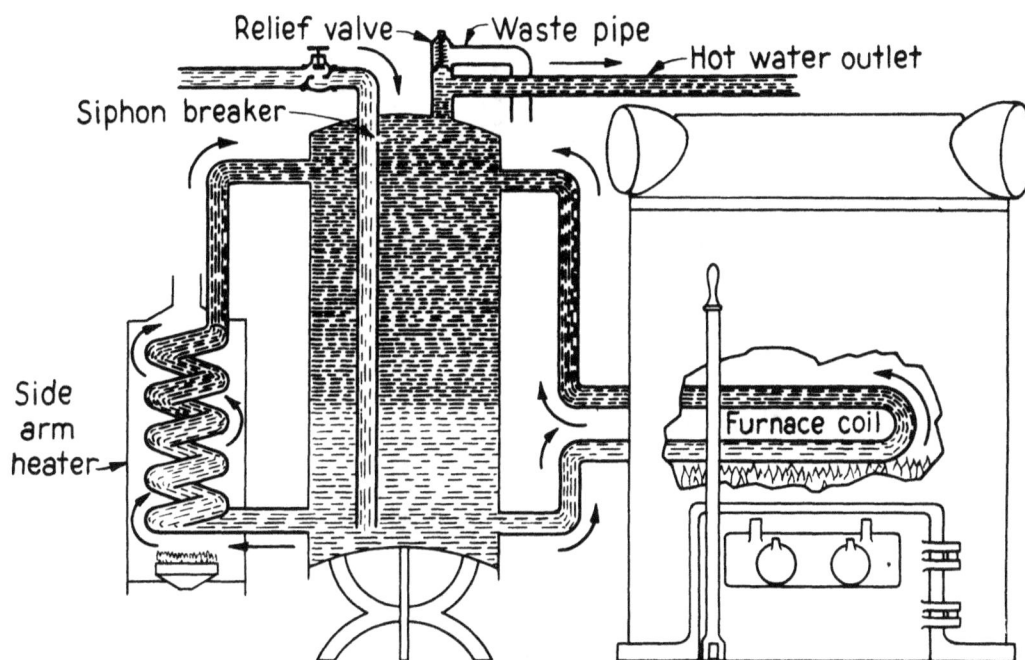

Figure 65. Common Range Boiler Installation Connected
with Furnace and Side Arm Heater

The tank is not usually insulated; however, it can be done profitably,
particularly if used with a heat source that is for heating water only.
When the heat source is a kitchen range or furnace coil it is sometimes
desirable for the tank to lose heat to keep the water from boiling.
Most automatic heaters are insulated.

Electric Water Heaters

Automatic electric water heaters are being accepted rapidly in rural area
where electric rates will permit. The advantages are that they are safe,

clean, fully automatic, can be installed almost anywhere, and require no firing."

The design is somewhat different from that of a range boiler. The cold water enters the bottom and is usually diverted by a plate or some means that will cause little commotion and hence will not disturb the hot water layer. A heating element, usually of 1500- to 5000-watt capacity, heats the water and sets up a circulation within the tank. The heated water collects in the top and a rather definite line is established between the hot and cold water layers. When the contents of the tank are entirely heated a thermostat disconnects the electric circuit. Some electric heaters are equipped with a second heating element and thermostat near the top of the tank, which provides a quicker source of hot water in case of heavy demands, and acts as a supplement for the lower unit.

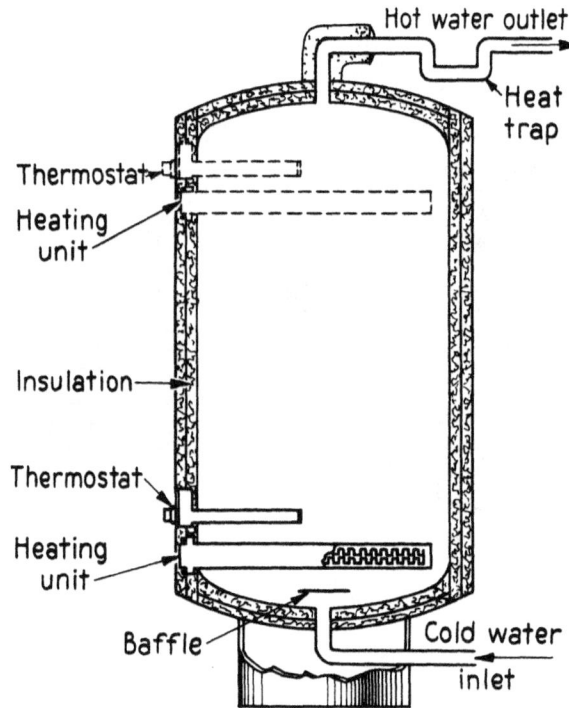

Figure 66. An Electric Water Heater

The siphon breaker arrangement used with the range boiler is not so badly needed since the heat concentration is not as great with an electric heater; the space around the element is large, allowing room for expansion, and, too, the heating element will "burn out" if it heats with no water in the tank.

Nearly all electric heaters are well insulated. Most manufacturers further conserve heat by installing a heat trap, either within the heater jacket or outside the jacket, as shown in figure 66. The purpose of the trap is to prevent hot water from circulating in a single pipe. The insertion of a U-shaped section in the discharge pipe causes the cold water to collect in the low section of the "U" and stops the circulation.

Electric water heaters for home use are available in almost any size from 1 gallon for sink installations to 150 gallons for large homes.

Miscellaneous Types of Electric Heaters

The side arm electric heater mentioned in connection with range boilers is widely used in some areas. Many were installed prior to the development of the heater just discussed. The efficiency is less and there is a greater tendency for the heater to lime than with the storage type.

Instantaneous electric heaters are available that attach to a faucet and heat the water as it is discharged. The wattage is high and the water flow so limited that they are normally not considered satisfactory.

A more popular unit is a small immersion heater that is dropped into a pan or kettle of water and connected to a convenience outlet. The heater is relatively fast for small quantities of water, but the heating unit is in direct contact with the water so that there is a decided risk of severe shock if a person touches the water and some grounded object at the same time. Immersion heaters that are well insulated can be secured but they are comparatively high priced and are not generally used.

Water Heater Installation

Location

A range boiler must be set close to the heat source, since the flow caus by the water being heated is relatively weak. This usually means that a range boiler will be set in the kitchen or basement close to a range or furnace. However, with a side arm heater as the only heat source, it ca as a last resort, be placed in the bathroom if space is limited. An un-insulated tank gives off considerable heat, which may be an advantage in winter, but a decided disadvantage in summer. Some fuels, particularly gasoline or kerosene, give off offensive odors and gas.

An electric heater is commonly placed in the basement or kitchen. Speci table top types are available for kitchen use. It can be installed in t bathroom, but it will not supply heat for warming the room.

If possible it is well to set the tank in the vicinity of the kitchen s: since the small, intermittent withdrawals are usually from there. To fi ther conserve hot water some plumbers use 1/2-inch pipe instead of 3/4-: between the heater and hot water outlets, unless the runs of pipe are qi long.

Provisions for Servicing

When hard water is heated it is occasionally necessary to replace the heater coils, or water front in the case of a kitchen range, because of lime deposited on the inner surface. The water flow is gradually restricted until there is little or no flow, and consequently limited hot water is available. When the coil is replaced the tank is drained, so there should be some provision for attaching a hose to a faucet at the base of the tank, unless the water can be discharged on the floor.

Automatic electric heaters are also subject to liming and occasionally a heating element will "burn out" because of a collection of lime on the sheath; consequently, they should be installed so they can be drained readily.

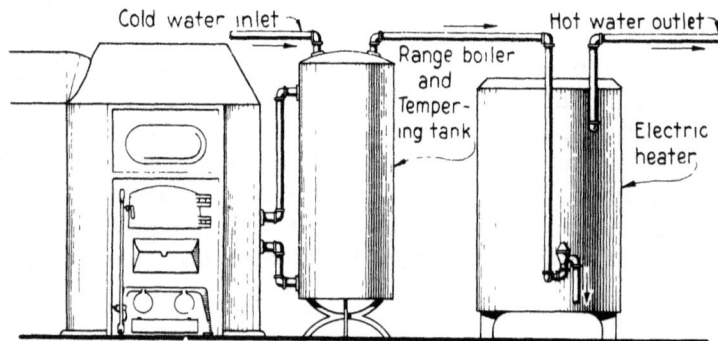

Figure 67. Range Boiler Used as a Heat Tempering Tank to Temper Water Entering Electric Heater

Relief Valve

A relief valve should be installed somewhere in the supply line, if a pressure system is used, to take care of water expansion or excessive heating of water. Such a condition often exists with a furnace coil in winter weather during heavy firing. Many farm installations are dependent on the pressure relief valve at the pump. There are two principal objections to this arrangement: (1) hot water backs into the cold water line, causing hot water to flow out of the cold water faucet when opened; (2) the type of relief valves generally used with pumps tend to stick and excessive pressures may be required to open them.

To avoid such possibilities it is good practice to install a relief valve near the heater. Several good relief valves are designed for the purpose

Operating Cost

Since the cost of heating water is dependent on temperature of incoming water, temperature to which water is heated, quantity of water used, degree of insulation, etc., it is difficult to arrive at a comparative cost for most farm families. With a hot water consumption of about 43 gallons per day, the following is a rough indication of cost:

1. A furnace water coil accounts for about 20 per cent of the total coal used by the furnace.
2. A laundry stove requires about 300 to 600 pounds of coal per month.
3. An automatic electric heater delivers about 3 gallons of hot water per kilowatt hour (water heated to about 150° F.).
4. Kerosene and gasoline heaters require 1 to 3 gallons of fuel per day.

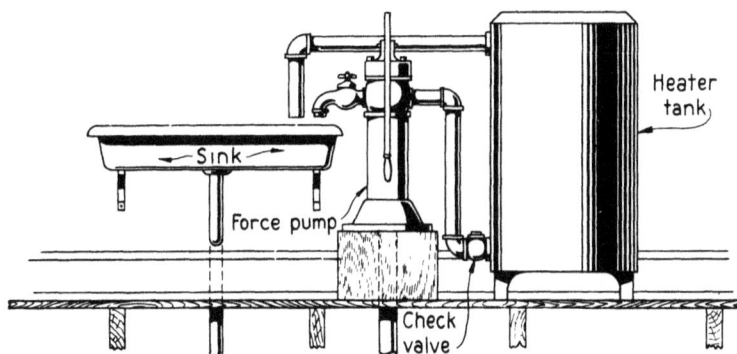

Figure 68. Arrangement for Using a Range Boiler or Electric Heater with a Hand Force Pump

WATER TREATMENT

Water Softening

Whether water is hard or soft is usually judged by how freely a lather is produced. Soft water requires very little soap, while hard water, depending on its degree of hardness, may require a considerable amount.

The principal disadvantages of hard water are:

1. Soap requirements are much greater.
2. Heating coils become filled with lime deposit.
3. A curd or scum forms on top of the water when soap is used and will lodge in the fiber of clothing, causing harshness.

4. Spots and streaks are left on dishes and glassware unless polished.
5. The skin becomes roughened from washing.
6. Iron in water will stain china and enamel fixtures.

Calcium and magnesium bicarbonates, calcium and magnesium sulphates, singly or in combinations, are the common causes of water hardness. Aluminum and iron also contribute to hardness but are usually present in such small quantities that their effect is negligible.

Various chemicals are available for adding to water to soften it, but probably the most satisfactory method for farm use is the zeolite softener that connects into the supply line of the plumbing system.

Zeolite is a name applied to a group of substances or minerals which have the capacity to gradually change composition in the presence of water without being dissolved themselves. By simply passing hard water through zeolite the substance takes up the calcium and magnesium and liberates sodium, which does not contribute to hardness itself. After the zeolite has taken up its capacity of calcium and magnesium the additional water that passes through it continues to be hard.

The remarkable feature of zeolite is that passing a concentrated solution of ordinary salt through the mineral will cause it to give up the calcium and magnesium and recharge itself with sodium from the salt, so that it again has capacity to soften water. Excess salt is then washed out of the zeolite. The softener is often connected to the hot water line only to reduce the total water passing through it, thus reducing the frequency of recharging.

Zeolite is available as green sand and white sand. The former is a natural mineral which is not as pure as white sand, but which has the ability to remove iron if the concentration does not exceed 4.5 parts per million.

White sand is synthetic and lacks iron removal capacity; however, it has at least four times the capacity of green sand to remove calcium and magnesium.

(Hardness of water is usually expressed in grains. However, it is sometimes referred to as parts per million. One grain equals 17.1 parts per million.)

Softeners are available with a zeolite tank only or with a second tank which contains salt brine for recharging. The latter type is usually worth the additional cost in added convenience (figure 69).

Entering salt brine into the zeolite is usually a matter of turning one valve, which siphons the brine into the zeolite. Excess salt is then washed out and drained away; then the main water supply is again reconnected and service established.

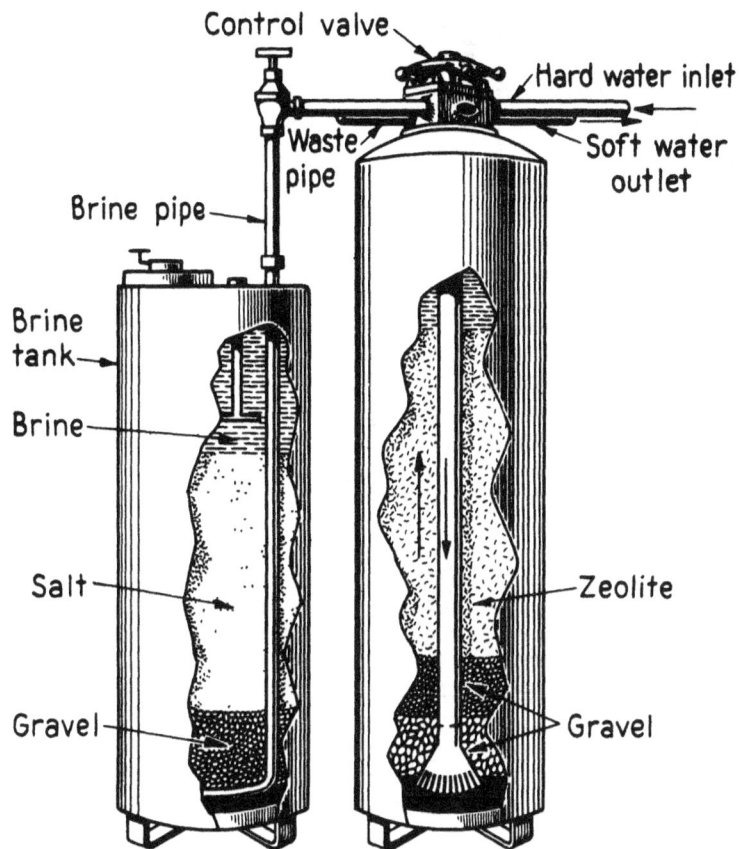

Figure 69. Zeolite Water Softener with Brine Tank

One common mistake is to purchase a softener that is too small. It should be large enough that a recharge is not necessary more frequently than every 4 to 6 weeks. A softener that is too small is usually neglected, resulting in part time hard water service.

Iron. Iron in well water is usually in a soluble form and can be removed by base exchange with green sand zeolite the same as calcium and magnesium. However, if it is exposed to air it will oxidize into iron in suspension and must be removed by filtering. The filtering can be accomplished in the zeolite softener, except that the flow of water must be down through the zeolite rather than up, as is the case with most softeners.

Sulphur. Sulphur can normally be removed by forced draft aeration. If aeration does not completely remove the sulphur it may be necessary to chlorinate the water.

Acid Water. Sometimes water is of an acid rather than alkaline nature, which may be the case with cistern water. If the acid is sufficient to require treatment it can be counteracted by running it through marble chip or limestone bed. The bed will be gradually used up and have to be replaced.

Commercial tanks for use with pressure systems are available.

Bad Taste. Fishy, earthy, marshy, and most other bad tastes, except those of sulphur and iron, can be removed by running the water through a charcoal filter bed.

Commercial tank units are available for pressure systems.

Chlorination. Often an unsafe but adequate water supply can be made safe by use of a chlorinator. Small chlorinators are available to be used in connection with a farm pumping system. The chlorine comes in powder form. It is mixed with water and the concentrated solution is injected by the motor driven chlorinator into the supply line in proportion to the water used. If a shallow well pump is used there are chlorinators that can be connected to the suction line and thus eliminate a motor (figure 70).

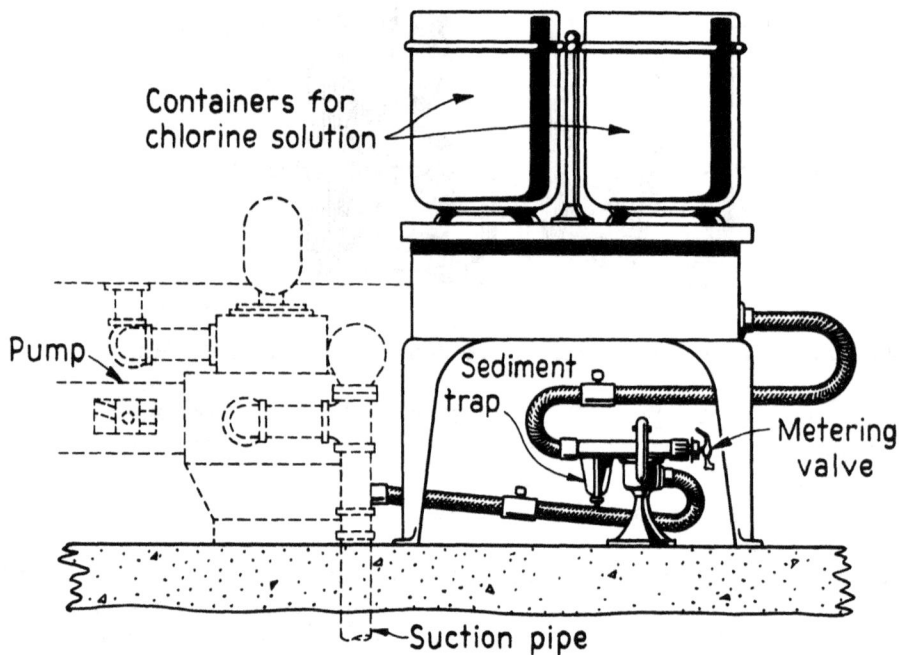

Figure 70. One Type of Hydro-Chlorinator for Questionable Water Supplies.

CHAPTER VIII

SEWAGE DISPOSAL

Sewage Composition

Fresh sewage is gray in color and has very much the appearance of soapy
dish water. For each 150 gallons of sewage there is about one pound of
solids. Of these solids about one half is mineral and the other half
vegetable and animal matter. The vegetable and animal matter breaks
down still further into about 40 per cent in suspension, which will settle out over a period of time, and 60 per cent in solution, which will
not settle out.

Disposal of the settleable and non-settleable portions and disposal of
the effluent water constitue the major problems of sewage treatment.

Septic Tanks

Sewage is usually delivered from the house to a special digestive tank
called a septic tank, or to a walled hole in the ground called a cesspool. Since the septic tank is recommended and most commonly used, it
will be discussed first.

Action within Septic Tank

A septic tank is simply a retaining chamber which assists and speeds the
natural decomposition processes. As the sewage enters the tank portions
of the solid matter settle to the bottom where decomposition begins. The
lighter particles and grease rise to the top to form scum.

The action of decomposition is brought about by certain types of bacteria
known as anaerobic bacteria that thrive in the absence of air. They at-
tack the organic solids and break them up into liquid and gases. The di-
gestion is almost complete with the exception of a small quantity of material which does not decompose and remains in the tank as sludge.

The bacterial action is essential to the successful operation of the
tank. When chemicals are emptied into the tank or discharged down the
drain so that they reach the tank, the bacterial action is retarded or
stopped; consequently it is not good practice to use lye and various
other commercial preparations to open clogged drains.

As the raw sewage enters the tank an equal volume of liquid is forced out
the discharge side, so that the liquid level remains the same in the tank

The inlet is provided with a baffle or other provision so that the incoming sewage is diverted downward. The arrangement permits the scum to remain undisturbed and limits the possibility of raw sewage being discharged immediately. A baffle at the outlet also helps accomplish the purpose.

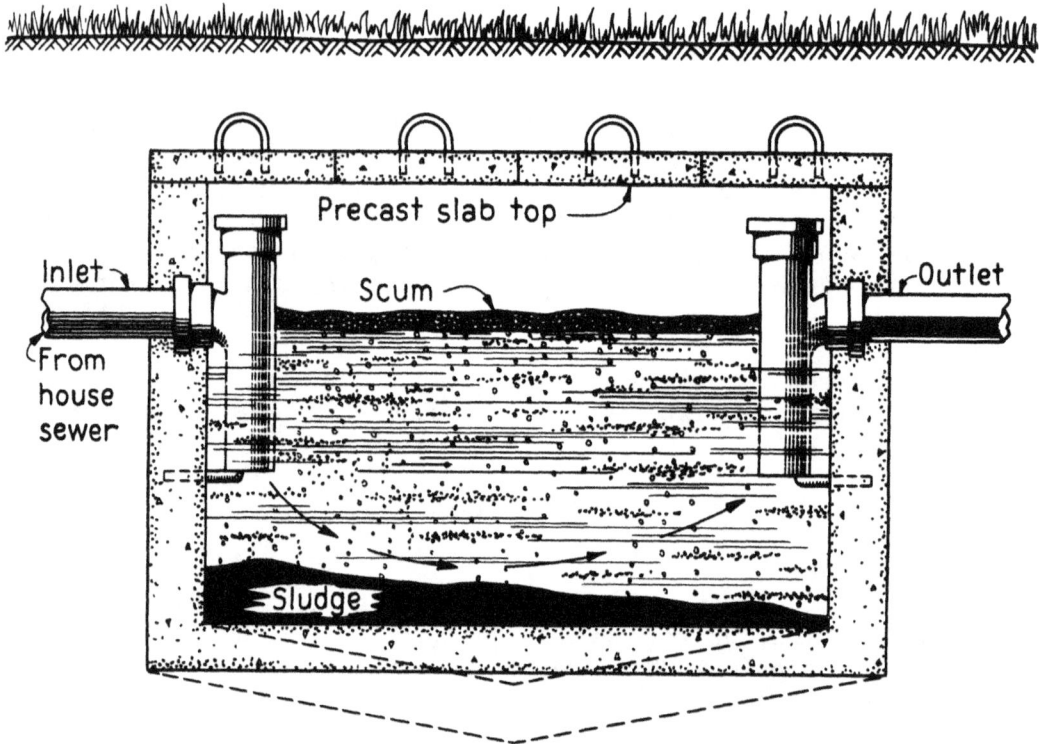

Figure 71.   Common Type Septic Tank. A Sloping Bottom
Aids in Removing Sludge.

Although the discharged effluent has most of the solid matter removed, it is far from pure and has an offensive odor.

Tank Size

It is generally thought that the tank should be large enough to accomodate the sewage discharged from the house over a 24-hour period; however, state recommendations vary from 8 to 48 hours. The water received in the tank is usually estimated to be around 30 to 50 gallons per person per day. Many states have established a minimum size for septic tanks, which varies roughly from 200 to 540 gallons capacity. Before a septic tank is install ed recommendations should be secured from the state health department. If there is any question as to whether a tank is large enough, be sure to use the next larger size. The mistake is using a tank that is too small.

In no case should water from rain spout be discharged into a septic tank.

Construction and Location

Septic tanks similar to the one shown in figure 71 are commonly used on farms; however, construction details vary with different states so that it is best to depend on the experience and judgment of the state health departments.

Iron, precast concrete, and vitrified clay tanks can be purchased, or a farmer may pour his own concrete tank. In any case the tank should be watertight, of heavy enough construction to be permanent, and amply large for present and future needs.

Iron tanks are rust resisting and heavily coated with a preservative. Opinion differs as to their permanence, some authorities claim that a 12-gage tank will last 10 to 15 years. Well-made concrete tanks should last indefinitely. However, a poorly constructed concrete tank may not last as long as a light weight iron tank.

In all cases the top should be provided with a removable cover for pumping sludge or checking on tank operation.

The tank may be located near the house, but if space is available it is well to locate it 50 to 100 feet away from a dwelling and at least 50 feet on the downhill side from the well. Many states require that the tank be at least 100 feet distant from a well. Its location and depth are often determined by the height of the lowest discharge in the house.

In most cases it is well to cover the tank with at least 12 inches of dirt to keep out flies, eliminate the possibility of offensive odors, and provide a more constant tank temperature to promote bacterial growth.

Care and Operation

An amply large, well constructed septic tank will need little attention. Removal of the sludge every 3 to 5 years is probably sufficient. In some cases a tank may not be pumped more than once in 10 years. On the other hand, small tanks may require cleaning yearly. (Sludge should not accumulate in excess of one-third of the tank depth.)

The sludge is usually removed with a diaphragm or pitcher pump and buried. If left exposed it will give off a disagreeable odor and may be a definite health hazard.

Sewer Line to Septic Tank

The main sewer line connects at the base of the stack and conducts the waste to the septic tank. The same type of cast iron pipe that is used in the house is used to lead to the septic tank, and equal care is taken to make certain the joints are watertight.

To avoid possible clogging, the main line should have a fall of about 1/4-inch to the foot. It is excellent practice to install a cleanout plug, as shown in figure 72.

Grease Trap

Grease from the kitchen sink sometimes causes clogging of the kitchen drain or main sewer line. It will also cause trouble in the septic tank by coating the small particles of sewage and making them buoyant and re-sistant to bacterial action. The undigested, greasecoated particles may then be carried out into the disposal field and may cause trouble there.

As shown in figure 72, a grease trap is a small tank into which the kitchen waste empties before it enters the main sewer line. Since grease is lighter than water it collects at the top and the water is drained from the bottom of the tank. The grease must be removed before the trap fills with it or the arrangement will be of no value.

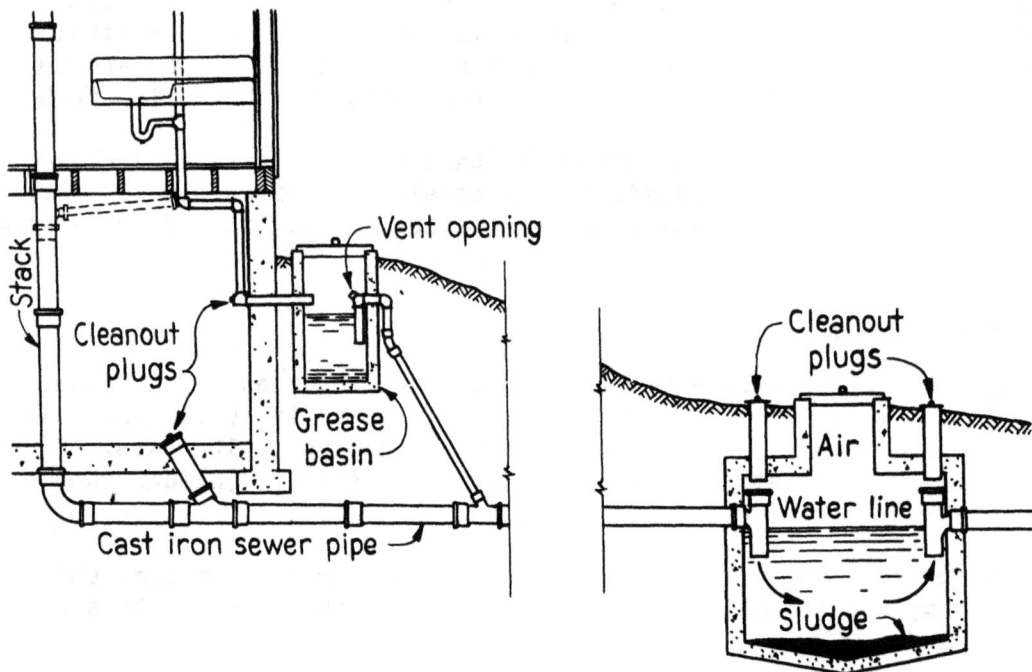

Figure 72. Connection between House and Septic Tank.
Note Grease Trap in Sink Drain Line.

Many farm installations operate successfully without a grease trap if care is taken to keep quantities of grease from draining down the sink. It is also possible, with the type of septic tank shown in figure 72, to remove some of the scum when the scum layer exceeds about 8 inches, and thus to some extent eliminate the necessity for a grease trap. However, the task is disagreeable and more likely to be neglected than if a grease trap is installed.

The grease trap may be built of concrete or masonry, or a metal tank can be purchased.

The trap should be installed just outside the house and as near the kitchen as practical.

Disposal Field

The disposal field usually consists of farm tile, laid in a long line, in several parallel lines, in S-shaped lines, or in some other arrangement that fits the topography and soil conditions. The purpose of the disposal field is to provide an outlet and distribution for the effluent

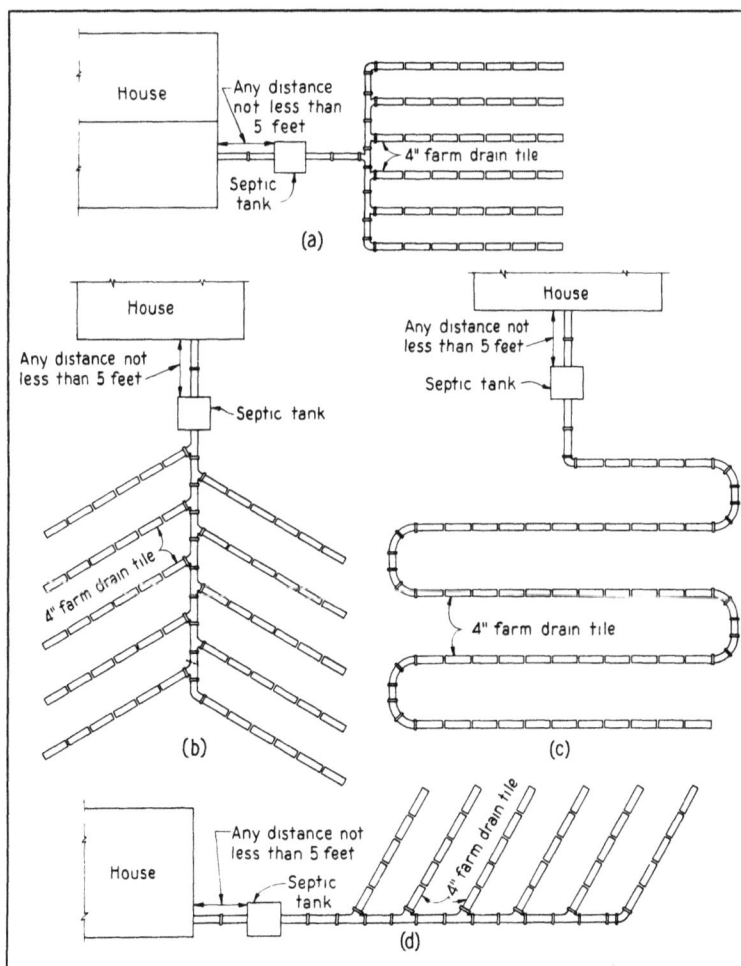

Figure 73. Types of Sewage Disposal Fields

It is permissible in some states to discharge the effluent into a lake or fast moving stream by special permission of the state health department. However, few farmers can take advantage of such conditions and must depend on a disposal field. Figure 73 shows various types of sewage disposal fields.

Since definite and detailed plans for installation are available from the various state health departments and agricultural extension services, our discussion will be confined to principles only.

Bacterial Action

The effluent, as discharged from the septic tank, is still far from pure, so that it is usually recommended that the disposal field be at least 100 feet removed from the well.

As the effluent is discharged from the septic tank and enters the tile line it is gradually absorbed from the open tile joints into the surround ing soil. Here in the presence of soil and air the nitrifying bacteria, which are found in the first few feet of top soil, reduce it to compounds necessary for plant life. As the water separates and filters through the soil it is gradually purified.

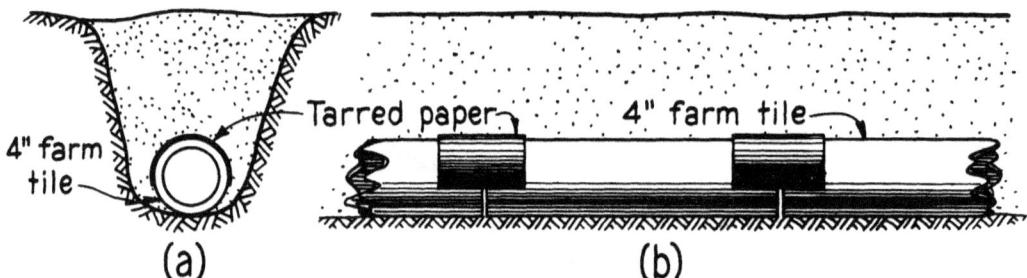

Figure 74. Method of Installing Disposal
Tile in Open Soils

Since loose soils absorb effluent quickly and tight clay soils are slow absorbers, the length and the grade of the tile line vary. For example, in loose soil the grade may be as much as 6 inches per 100 feet, so that the effluent will flow rapidly and not be absorbed in the first few feet. Under such conditions the disposal field may be figured at the rate of 30 feet per person.

In contrast, under similar conditions but with a clay soil, the grade may be 2 inches per 100 feet and the disposal tile length figured at the rate of 50 feet or more per person.

For tight soils it is often recommended that the tile be laid on a bed of crushed rock, as shown in figure 75. In this manner the total absorption area is increased.

In case the latter method is not adequate to take care of the effluent,
a subsurface drain may be used in connection with the disposal field
line, as shown in figure 76. Bacteria act on the effluent as it passes
through the filter. The subsurface drain then discharges the effluent
to a small stream, to a storm drain, or in some cases to the ground surface, provided it is well removed from buildings and wells. However,
some health departments feel the safest practice is to extend the subsurface drain sufficiently that the soil will absorb all the effluent.

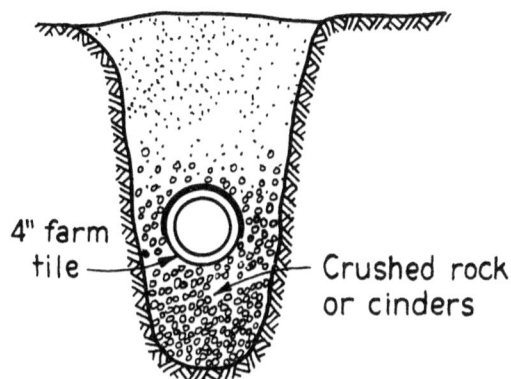

Figure 70

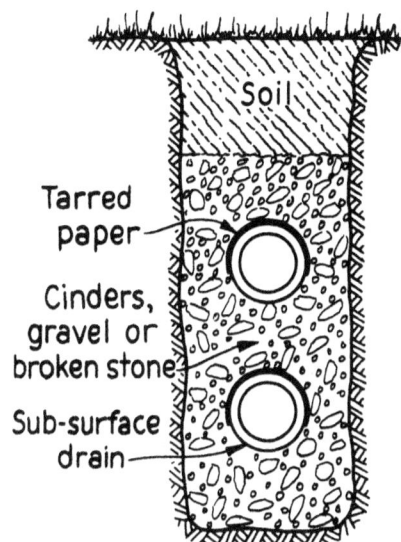

Figure 75. Tile and Gravel Arrangement for Tight Soils

Figure 76. Arrangement For
Very Tight Soils

It is essential that little or no grease enter the disposal tile. Once
it has been absorbed into the soil the spaces between the soil particles
become filled, gradually reducing the capacity of the soil to absorb effluent and limiting the effectiveness of bacterial action.

Cesspools

Cesspools were one of the early forms of sewage disposal on farms. They
are still used to some extent, but it is commonly recognized that they
are a menace to health and should be prohibited.

When a cesspool is used it means that raw sewage is dumped directly into
a rather large, deep, loosely walled hole. The water and small particles
of raw and digested sewage are gradually absorbed into the soil. The
soil becomes saturated so that anaerobic bacteria are the only active
type. The sewage is often below the level of nitrifying bacteria. As
grease is absorbed in the ground and residue accumulates in the cesspool,
it has less and less capacity so that it becomes necessary to construct
another.

Often the raw sewage finds its way into water-bearing strata where it may
become a menace to a whole community. In fact, there have been cases of
old abandoned wells being used as cesspools.

CHAPTER IX

FIXTURES AND FITTINGS

Few farm families can purchase plumbing fixtures with a feeling of certainty. There are problems of quality, design, color, finish, durability, sanitation, etc., that should be considered along with cost. Cost is necessarily a factor; yet, with a definite sum to spend for plumbing equipment, it is often possible to select items that cost little or no more that will fit farm conditions to a much better advantage.

Construction Materials

Sinks, lavatories, bathtubs, closets, and sometimes laundry tubs constitute the plumbing fixtures for farm homes. Of this group the first three --sinks, lavatories, and tubs--are commonly made of cast iron. However, more recently light weight types made of ingot iron and pressed steel have entered the field, and vitreous china lavatories have become more popular.

Porcelain Enamel

In order to secure a smooth, easily cleaned surface porcelain enamel is fused on iron or steel. Porcelain enamel is glass and is in no way related to enamel paints except in appearance; it is smooth and glossy. The fixture is normally sprayed with a ground coat, which is baked on and helps fuse the porcelain to the metal. The ground coat is followed with porcelain coats which build up to about 1/16-inch thickness. Each porcelain coat is baked on at temperatures of 1350° F. or higher, in order to make the porcelain fuse tightly to the metal.

A good porcelain job is important to the life and appearance of the fixture. Most companies make every effort to eliminate defective fixtures before they reach the customer. In spite of this effort, a fixture may prove defective after it has been installed, so that doing business with a reputable dealer is about the only assurance a purchaser has that the equipment is of first quality.

Since porcelain is glass, it will crack or chip off from a blow or the impact from a falling object. It will also scratch easily. If acids are present, such as are found in the juice of grapes and citrus fruits, ordinary porcelain will cloud and roughen.

The latter condition is overcome by using acid-resisting porcelain, which is a porcelain of about 95 per cent silica content as compared with about 45 per cent in ordinary porcelain. It is impossible to tell the difference between the two porcelains by their appearance, so that the dealer and manufacturer are again the only assurance the purchaser has as to which type is being purchased.

Colors

Colored fixtures usually cost 10 to 20 per cent more than plain white fixtures. Although there are many shades with commercial names, in general the colors are green, orchid, ivory, blue, light brown, and black.

Vitreous China

Closet bowls, closet tanks, and some of the higher priced lavatories are of vitreous china. Vitreous china is molded clay, baked at a high temperature. To give it a glossy finish it is dipped in glaze and baked again. The same colors used with procelain enamel are added in the glaze coat. The resulting product is heavy, smooth, glossy, durable, and acid resistant.

Lavatories of china cost about 15 per cent more than enameled iron.

Fixtures

Bathtubs

Bathtubs are of three general types:

    1. Built-in
        (a) Recessed type
        (b) Corner type
    2. On legs
    3. On base

The various types are available in a wide range of sizes, varying in length from 4 to 6 feet, in width from 26 to 36 inches, and in height from 16 to 22 inches. Square tubs are available that range from 43 to 48 inches. All dimensions are outside, overall measurements.

Figure 77. Built-in Type
Tub with Shower

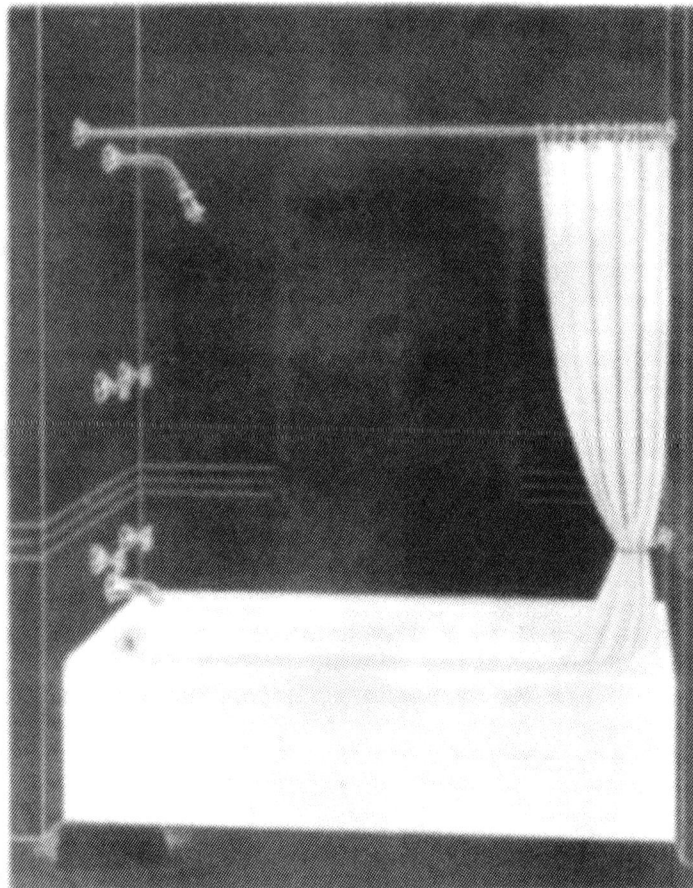

Built-in Tubs. Both recessed and corner built-in tubs are popular.
Although they are somewhat more expensive and should be used with a
moisture resistant wall bordering the tub, they have the following
advantages:

1. They are pleasing in appearance.
2. The problem of cleaning and painting the floor under
    the tub and side walls adjacent to the tub is
    eliminated.
3. They fit into a more limited space than other types.
4. They are more adaptable to shower installations.

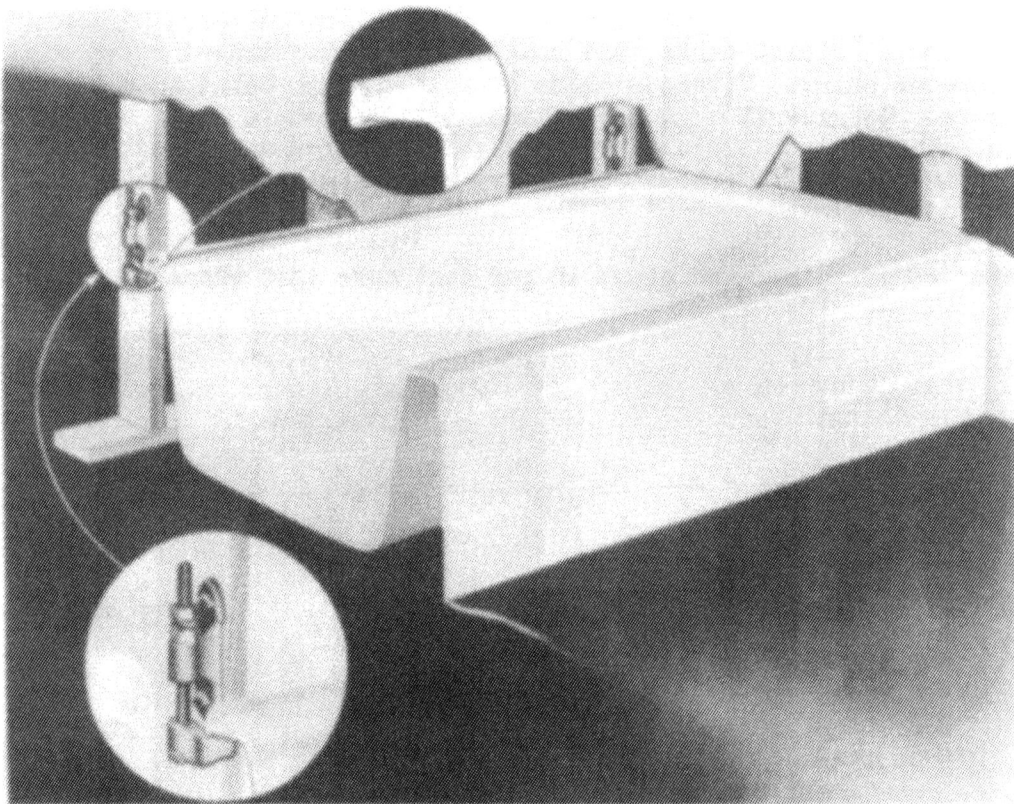

Figure 78. Adjustable Tub Support

There is a disadvantage in some cases. Because of the shrinking of joists
or settling of the house a crack develops between the tub and wall. This
can be overcome by using the adjustable metal support shown in figure 78.

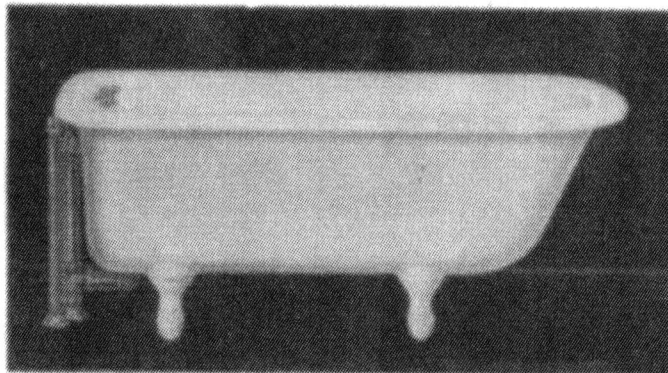

Figure 79. Tub on Legs

Tub on Legs. Figure 79 shows the tub on legs, the least expensive type. It is not as attractive as the built-in tub, it is more difficult to clean and paint around, and the piping is usually exposed. It is not well adapted to shower installations. However, it stands several inches above the floor, which requires less stooping to clean the inside of the tub, and it will give as much service as other tubs.

Tub on Base. The principal difference between the tub on base and the preceding type is the solid base on which the tub is mounted. It is sometimes offered at a low price with paint enamel finish inside. Its acceptance on farms has been limited.

Lavatories

Lavatories are of three general types:

1. Wall hung, available without legs, with legs, and bracket mounted (figure 80 a, b, c).
2. Pedestal (figure 81a)
3. Cabinet (figure 81b)

Standard lavatory height is 31 inches. As will be noted later, lavatories are made in a wide range of sizes. Unless space is a limiting factor, the larger sizes are much more satisfactory, because of a large bowl, a wide rim on which to set toilet articles, and a raised edge around the rim to keep the water from spilling over the side.

Wall hung Lavatory. The common wall hung type without legs or brackets is the most popular. It is relatively inexpensive, requires no floor space, and can be solidly mounted. The same is true of bracket mounted fixtures, except that they are sometimes more costly. These two types range in size from 18 to 26 inches in width and from 14 to 21 inches from front to back.

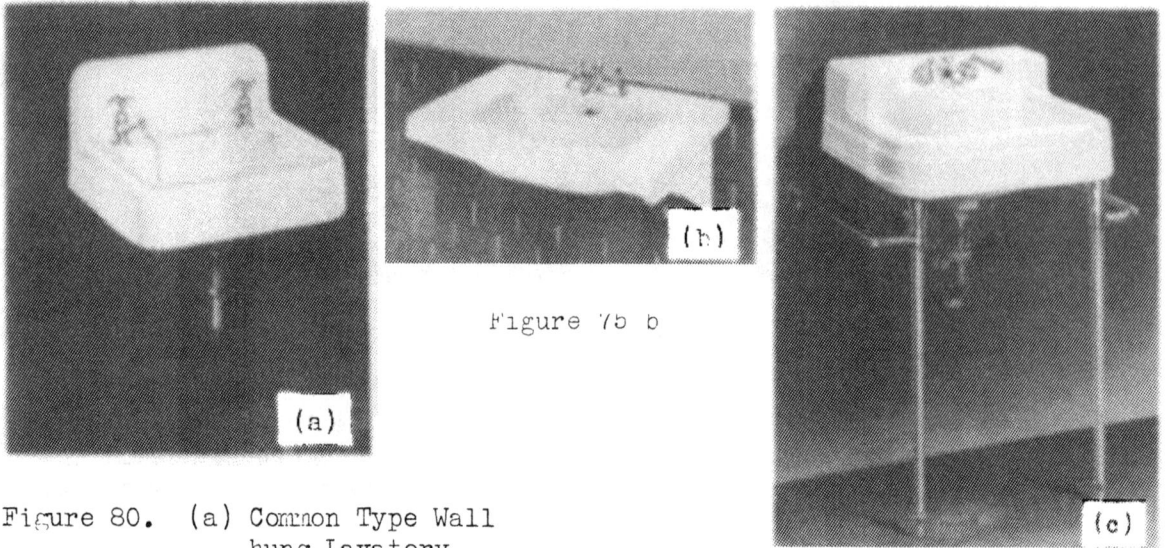

Figure 80. (a) Common Type Wall
                hung Lavatory
            (b) Bracket Mounted
            (c) Leg Type

The leg type is more expensive but provides additional support for the front of the lavatory. In most cases, however, the legs are added for appearance rather than for support. The size range is 20 to 27 inches in width and 14 to 22 inches from front to back. (Some types are made that are completely supported by legs.)

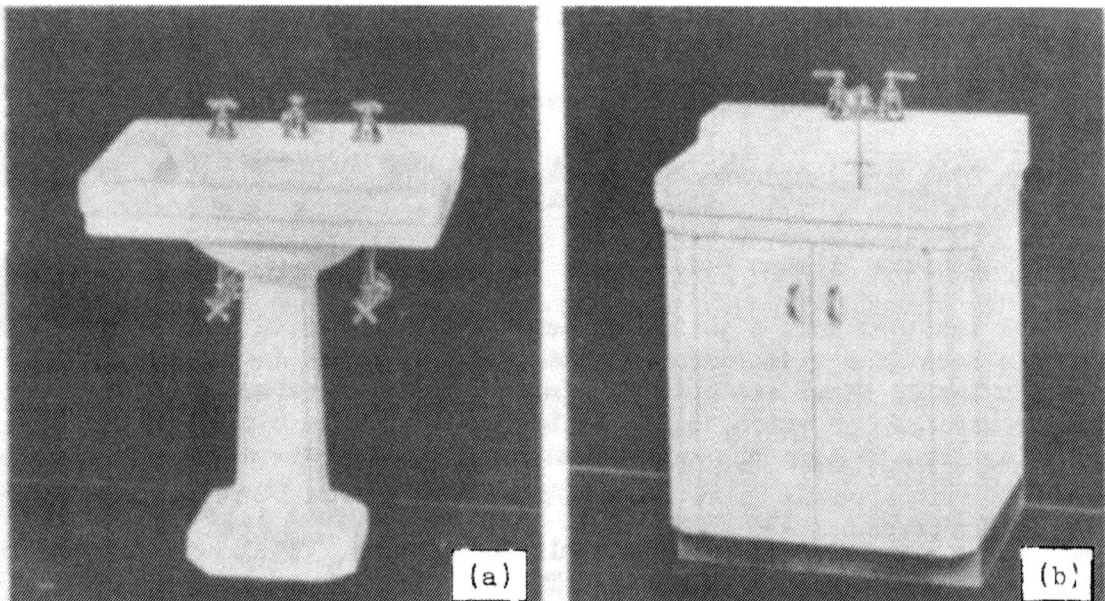

Figure 81. (a) Pedestal Lavatory
             (b) Cabinet Type

Pedestal Lavatory. The pedestal lavatory is available with larger overall dimensions than the wall hung type. In width they range from 24 to
30 inches and in depth from 20 to 24 inches. The pedestal supports the
lavatory so that it stands about 2 inches away from the wall. It has
the advantages of being impressive in appearance and being provided with
a wide ledge around the bowl. It has the disadvantage of being slightly
harder to service and, since it occupies floor space, it is an extra
item to clean around. Initial cost is considerably more than for most
wall hung types.

Lavatory on Cabinet. The cabinet types are smaller than pedestal types.
They are commonly available in widths of 19 to 24 inches and in depths of
17 to 20 inches. What has been said of pedestal lavatories applies in
general to the cabinet type, except that it has the added advantage of
storage space in the cabinet. The cabinet is usually of enameled steel.

Closets

As mentioned previously, closets, including tanks and bowls, are of vitreous china. They are available in a wide variety of plain and modern-
istic designs and in colors the same as those for other bathroom fixtures

As shown in figures 82 to 85, the residential type of closet bowl operates on a siphon principle. However, there are variations in the design
to secure certain desirable features.

For residential use closet bowls can be classified into three general
types:

    1. Washdown bowl
    2. Reverse trap bowl
    3. Siphon jet bowl

Figure 82. Common Washdown
Closet Bowl

Figure 83. Washdown Closet
with Jet

Washdown Bowl. The washdown bowl is simple, efficient within its limitations, and least costly. The trapway is at the front and is somewhat smaller than in other bowls, since proper functioning of the bowl is dependent on siphon action only. For farm homes without adequate heat in the bathroom, where there is a possibility of the bowl freezing and bursting, the common washdown type has special merit in that the water can be easily swabbed out.

Some types of washdown bowls are provided with a jet which aids in making the flush action more positive (figure 83). When the bowl is flushed a small stream of water spurts from the jet into the upper arm of the trap and starts the siphonic action.

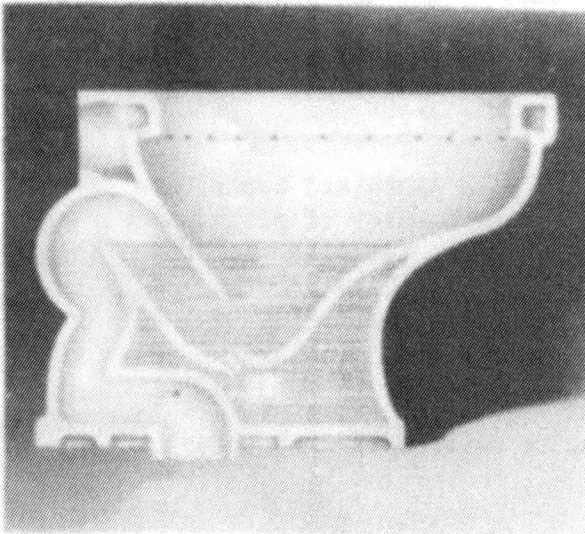

Reverse Trap Bowl. The reverse trap bowl is very similar to the washdown bowl, except that the trapway is placed at the rear, making it suitable for elongated rim construction. It is available with and without a jet. The water surface is larger, the water seal deeper, and the action of the bowl is more quiet and efficient than the washdown type. It is usually a little higher priced (figure 84).

Figure 84. Reverse Trap Closet Bowl

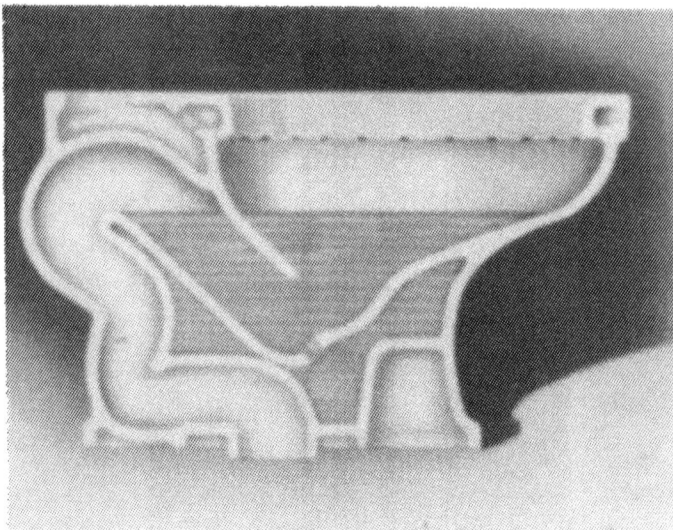

Siphon Jet Bowl. The siphon jet bowl is similar in appearance to the reverse trap bowl. However, the trapway is larger and the water seal deeper. A jet is provided in the bottom (figure 85).

The design permits a minimum amount of fouling surface and makes the bowl quiet and positive in operation.

It is the highest priced of the three types.

Figure 85. Siphon Jet Closet Bowl

<u>Tanks</u>. The tanks are very close to the bowls on modern closets, vary-
ing from a few inches above the bowl to sitting directly on the rear of
the bowl. The arrangement contributes to quiet operation.

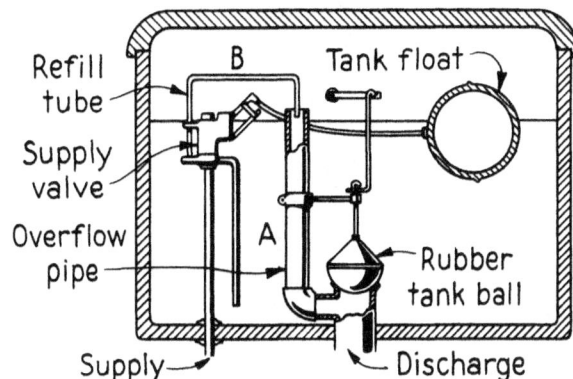

Figure 86. Flushing Mechanism in a Closet Tank

Operation is relatively simple. When the tank is flushed a series of levers raise the rubber ball valve on the discharge. The ball is buoyant and
tends to float until the water level recedes to a point where the rubber
ball is drawn back in position by suction and held in place by the water
pressure above it.

As the water level recedes, the metal float drops, opening the supply
valve, which allows water to enter through supply pipe A to fill the tank
(figure 86). A small amount enters through pipe B into the overflow pipe
to reestablish the seal in the closet trap. As the water level rises to
the proper position, the metal float closes the supply valve.

Showers

Showers are becoming more and more popular on farms, particularly with
men folk. The desire for some kind of shower is evident in the makeshift
arrangement found in the outbuildings on some farms without water service, which consists of a small elevated container with a shower head.

The advantages of a shower are that (1) it can be low in cost, depending
on type; (2) little water is required; (3) it is immediately available;
(4) it is fundamentally clean, since clean water is continuously sprayed
on the body; and (5) it is stimulating.

To get full year-round benefit and enjoyment from a shower it should be
installed where there is ample light, ventilation, hot water, and room
heat.

<u>Tub-Shower Combination</u>. The mounting of a shower head over a built-in
bathtub is probably the most satisfactory arrangement, because it best

Figure 87. Shower with Ring Curtain for use with Tub on Legs

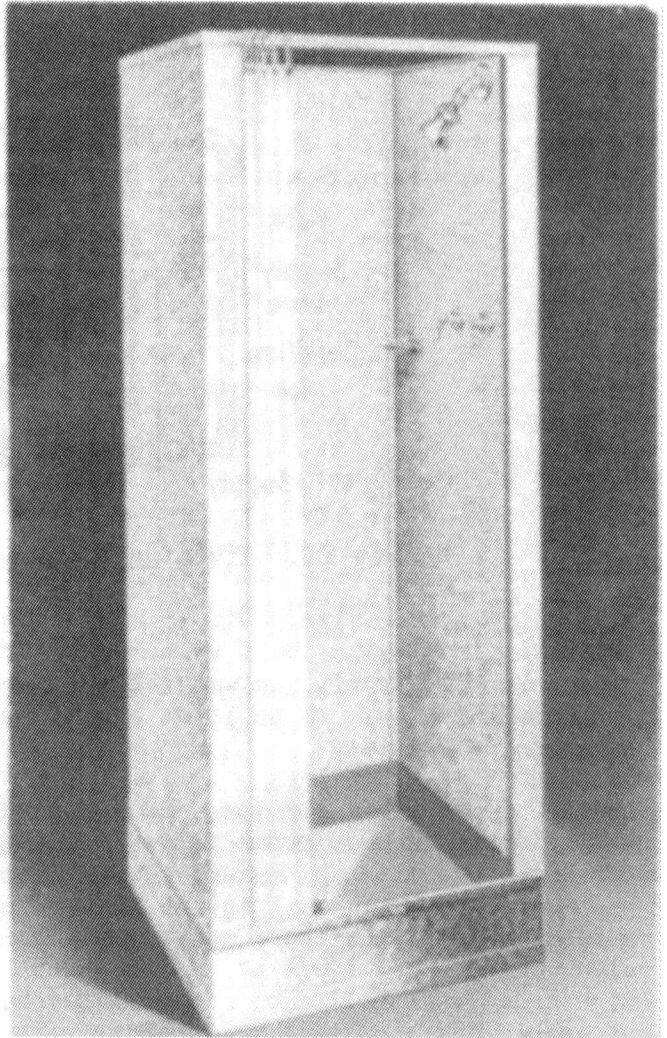

Figure 88. Commercial Type Shower Stall

meets the conditions just mentioned and requires no additional space (figure 77). The walls adjoining the tub should be waterproof up to a height of about 6 feet. By mounting a shower curtain over the rim of the tub, and suspending it from a rod about 6-1/2 feet from the floor, the spray can be easily confined. The tub then acts as a receptor and drain.

A less satisfactory arrangement is shown in figure 87. A rubber hose or pipe connects with the bathtub faucet to feed the shower head. It can be used with a leg type tub, but it has the disadvantage of being rather unattractive and providing too limited a space for the occupant.

Shower Compartment. A shower compartment is a stall about 32 to 36 inches square and 6 to 7 feet high that can be used in place of a tub. It has the advantage of doubling the bathing facilities of a home if it is located in some other place than the bathroom, an arrangement particularly desirable for large families.

Well built commercial or home constructed types that are watertight and
easily kept clean are desirable if ample light and ventilation are provided. Too often a shower compartment is constructed of wood panels and
located in a remote, poorly ventilated corner of the basement. The moisture is retained several hours after a shower so that the wood decays,
creating odors and a harbor for insects.

A much better low cost arrangement is to provide a shower head in the
basement without a stall but with shower curtains to form a temporary
stall when needed. When not in use, it can be easily ventilated and
cleaned.

Shower Head. The shower head should be set at an angle so that the stream
will be directed from the side rather than from above. The arrangement
avoids hair wetting and permits the shower head to drain easily after use.

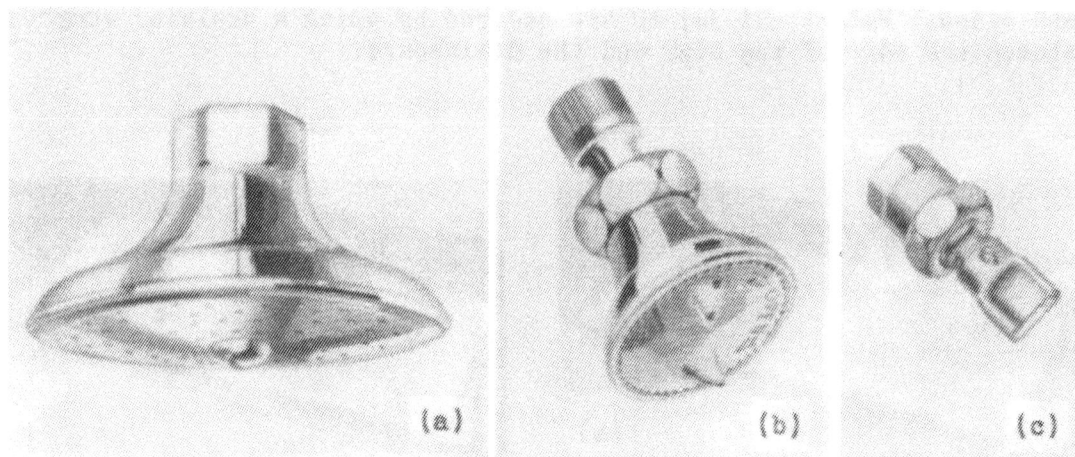

Figure 89. Shower Heads
(a) Rain Head
(b) Circular Spray
(c) Economy Head

Common types of shower heads can be classified roughly as follows:

1. Rain head--a relatively wide face plate with perforations
    evenly distributed (figure 89a).
2. Circular spray--grooves around edge of face plate. More
    economical with water than rain head (figure 89b).
3. Economy head--a restricted nozzle type that gives finer
    spray, limits the water spread, and is more economical
    with the water than the other two types (figure 89c).

The shower head should be designed for easy cleaning and adjustment to
different angles.

Sinks

A sink is usually the first plumbing fixture to be installed in the farm home. It can be inexpensive and easily installed, yet many farm homes are still without this simple convenience.

Sinks are usually of porcelain enamel on cast iron, ingot iron, or pressed steel; however, some of the more expensive types are of stainless steel and monel metal. The latter types are rust proof, are unaffected by acids do not scratch or dent easily, and resemble nickel in appearance. Porcelain enamel types are available in both plain and acid resisting finishes, the latter being most practical. Both finishes are available in colors.

Sinks can be classified roughly into flat rim and roll rim types.

Flat Rim Sinks. Flat rim sinks are less expensive, ranging in price from about $3 up. Inexpensive installations can be bracket mounted or they can easily be placed in a built-in cabinet with a drainboard on either or both sides. Watertight joints are secured by using a caulking compound between the edge of the sink and the drainboard.

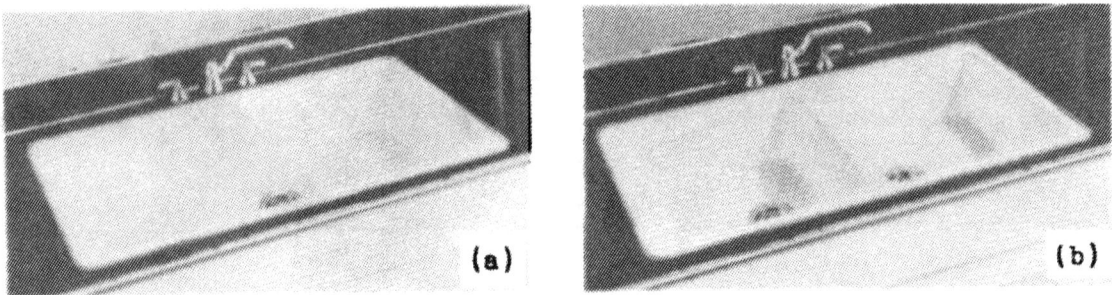

Figure 90. Flat Rim Sinks
    (a) Single Compartment Type
    (b) Double Compartment Type

Some flat rim sinks are built with drainboards. If desired, a simple sink installation without a cabinet can be equipped with a supplemental drainboard, as shown in figure 91a.

Sizes vary for single compartment sinks from about 12 by 18 inches to 36 by 20 inches and from 5 to 8 inches in depth. Double compartment sinks are also available (figure 90b). Larger size sinks are preferable.

Flat rim sinks have become more popular since the demand of housewives is for long, unbroken work surfaces in the kitchen. Such a surface is accomplished with a cabinet top in which the sink is recessed and the work surface covered with linoleum or some other waterproof covering.

If the drainboard is constructed of pressed wood or cypress, a covering is not essential except for appearances.

Roll Rim Sinks. Roll rim sinks are of the type shown in figure 91. On some the rim is extended down to form an apron. All have backs of from 4 to 8 inches. They are also available with single and double compartments, right or left hand drainboards, double drainboards, and no drainboards. They can be wall hung or cabinet mounted.

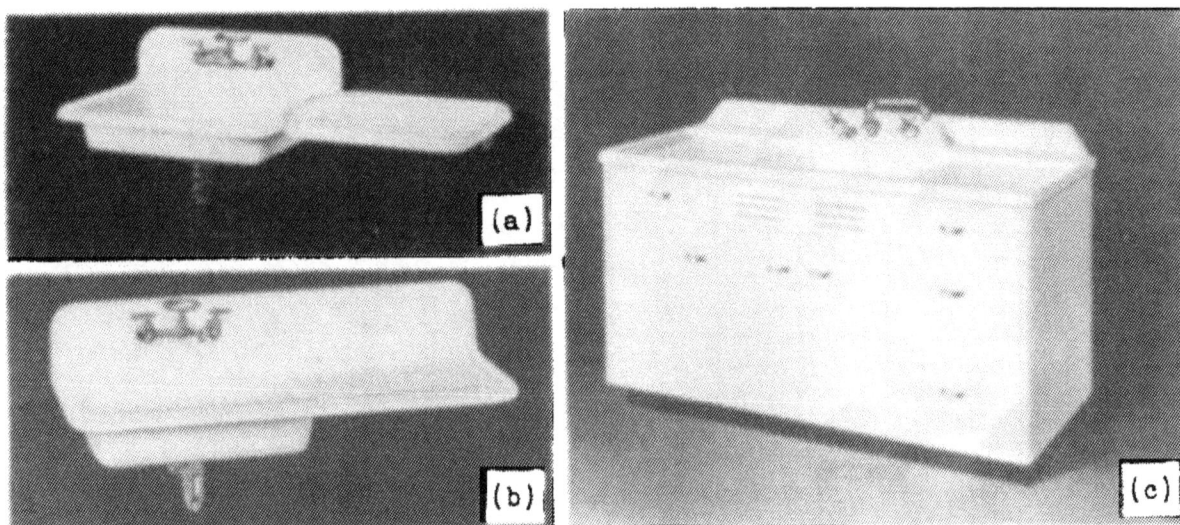

Figure 91.   Roll Rim Sinks
       (a) High Back Type with Supplmental
           Drainboard
       (b) Single Drainboard Type
       (c) Cabinet Type

Drainboards are generally about 18 to 24 inches long and provided with a raised rim to keep water from spilling over.

Laundry Tubs

With the wide use of inexpensive portable laundry tubs on farms the permanently installed type is slow in being accepted.

Permanently installed tubs range in price from $10 up, depending on material and design. They are commonly made of concrete, vitreous china, enameled cast iron, or enameled steel, and in single or double compartments.

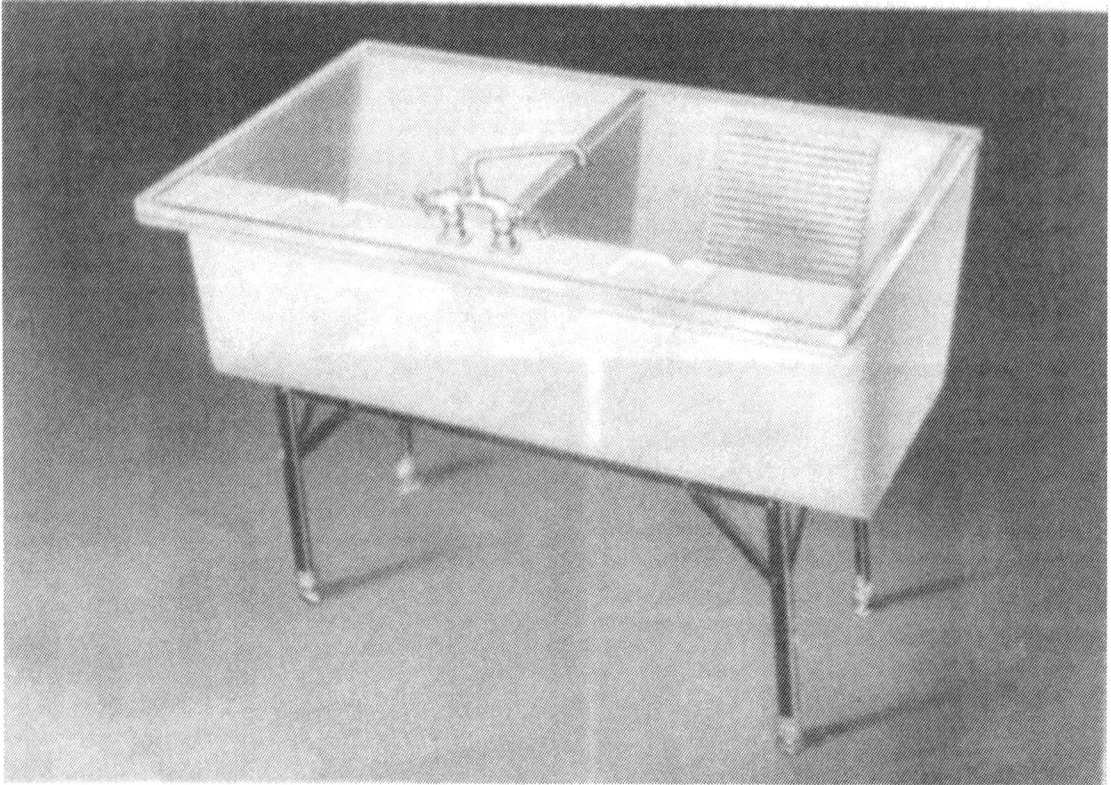

Figure 92. Laundry Tub

Fittings

Such items as faucets, wastes, etc., used with sanitary fixtures, are
of brass and are almost always chromium plated. Earlier fittings were
nickel plated and porcelain handles were used to some extent. Nickel
plated fittings are still available, but porcelain handles have been
discontinued.

In general, better quality fittings are heavier, which means longer
life due to heavier construction throughout.

Valves

Nearly all faucets used as fittings on present plumbing fixtures are of
the compression type shown in figure 93. Inexpensive fittings consist
of two separate valves--one for cold water and the other for hot water.
However, for baths, showers, and laundry tubs, and to a great extent for
lavatories and sinks, the two valves are designed with a common outlet
and thus become a double valve. The arrangement is a distinct advantage

in mixing hot and cold water to secure the desired temperature. These valves are available for concealed or surface installations, and for vertical, horizontal, and angle mounting.

A notable feature on modern fixtures is that the water outlet end of the spout is always above the rim of the fixture. This feature is a sanitary measure to prevent possible siphoning of used water from a filled lavatory bowl back into the supply pipes in case the supply lines are disconnected and a faucet opened. Many modern valves are made with removable seats, which are desirable when repairs are necessary.

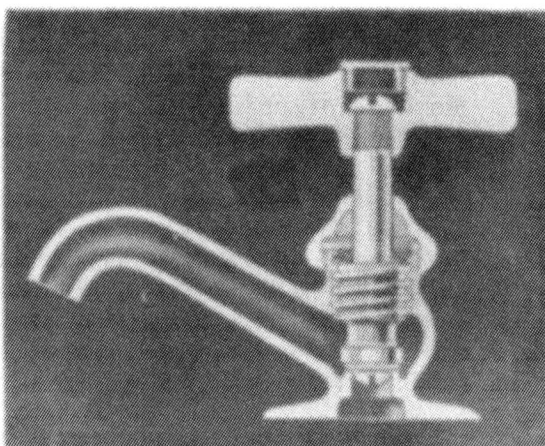

Figure 93. Compression Type Valve

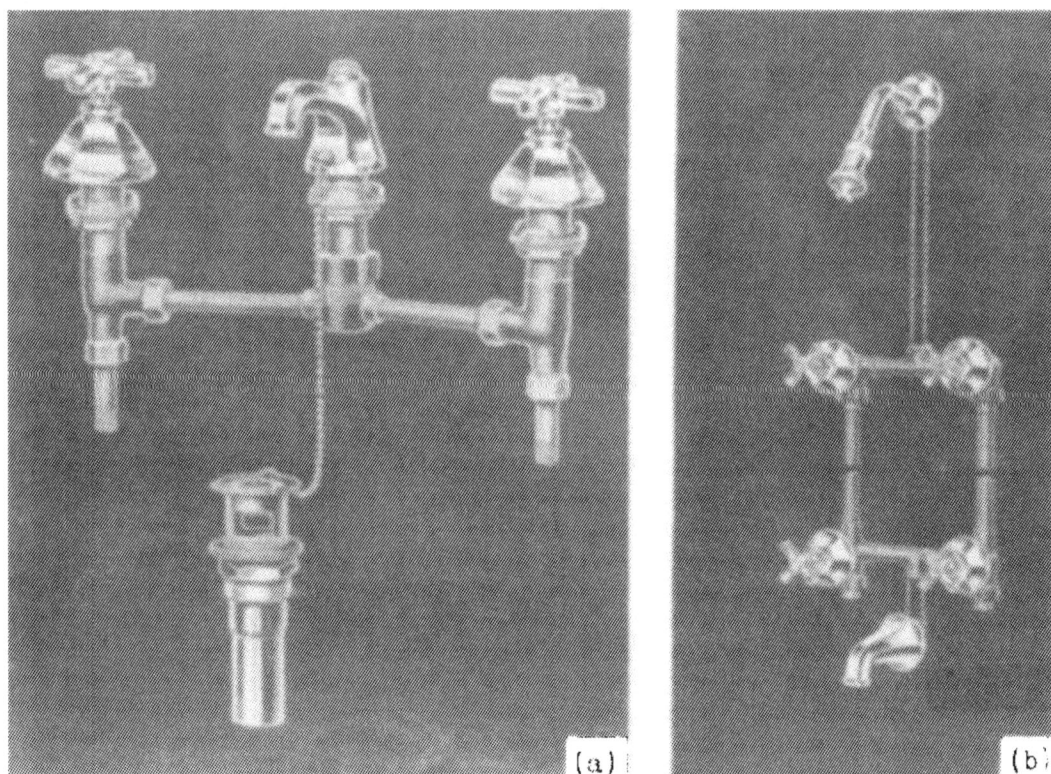

Figure 94. (a) Double Faucet
(b) Bath and Shower Combination

Waste Fittings

Lavatories and Bathtubs. Three types of waste fittings are used for
lavatories and bathtubs:

1. Pop-up type waste valve with overflow opening in upper part
    of fixture, which is expensive but is considered most
    sanitary of the three types (figure 95a).
2. Removable rubber stopper with overflow located in upper
    part of fixture. Least expensive (figure 95b).
3. Concealed waste valve and overflow (sometimes called stand-
    ing waste and overflow). This type has gained disfavor
    because of the unsanitary condition created by the inter-
    change of water between the overflow riser and the fixture
    proper when the fixture is in use. Sediment from the over-
    flow often works back into the bowl or tub (figure 95c).

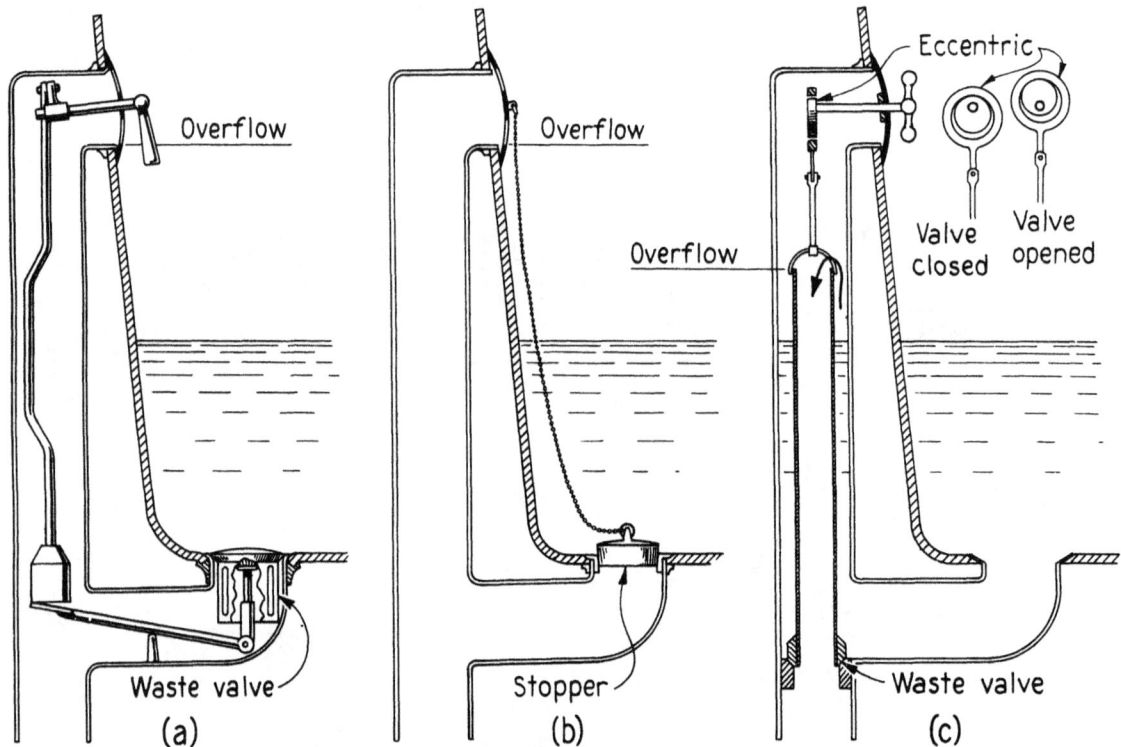

Figure 95.   (a) Pop-up Type Waste
              (b) Stopper Type
              (c) Concealed Type

<u>Sinks</u>. There are two types of sink waste fittings:

    1. Plain strainer type (figure 96a).
    2. Basket Strainer type (figure 96b).

The former is less expensive but the basket strainer type is generally preferred. It has the advantage of catching large particles of waste below sink level. For emptying, the basket is lifted out. Many fittings of this type are provided with a stopper so that the sink can be used as the container for washing dishes.

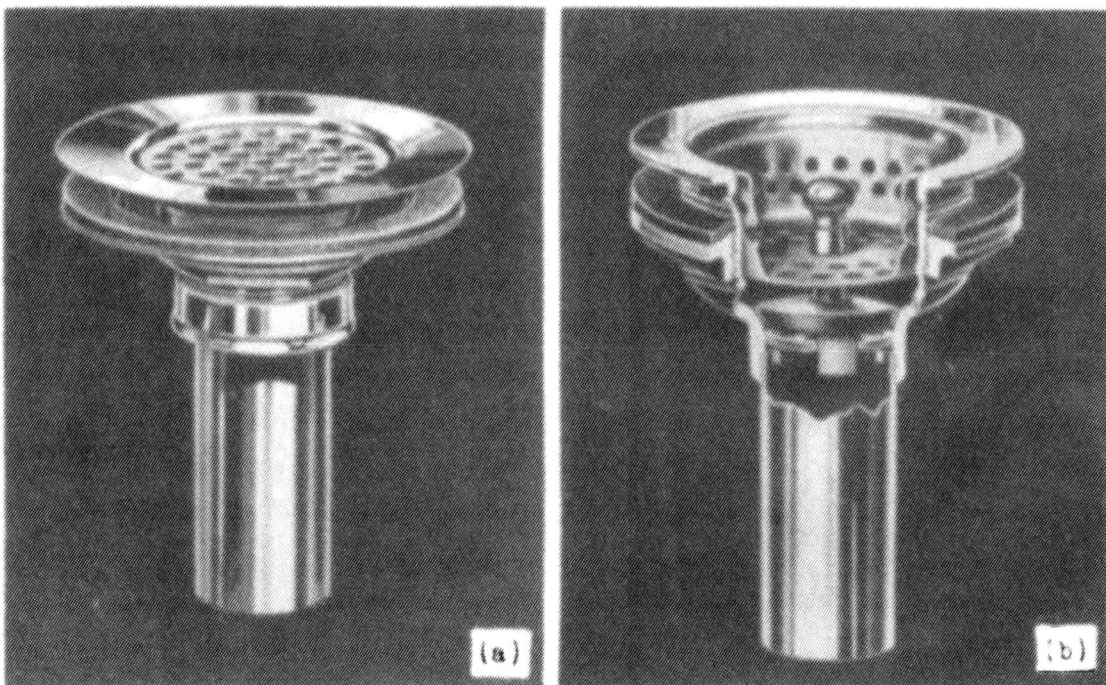

Figure 96. Wastes for Kitchen Sinks
         (a) Plain Strainer Type
         (b) Basket Strainer

CHAPTER X

BATHROOM PLANNING

The installation of a bathroom in a farm home is a step of vital impor-
tance. It is the type of improvement that is permanent and involves
considerable cost; consequently it should be well planned. Too often
this is not done.

It is not uncommon to find farm homes with a portion of an open porch
converted into a bathroom, with the only door opening directly onto the
exposed porch. In other cases the bathroom is located so that it is
necessary to go through a bedroom to reach it. Still more common is a
bathroom with two or three doors, some of which some member of the family unthinkingly leaves locked after finishing in the bathroom.

Although it is probably impossible to locate and design a bathroom with
all the good features contained in this discussion, by planning it is
usually possible to eliminate the conditions just mentioned, as well as
many others.

Installation Cost Factor

The cost of a bathroom installation varies with each individual case;
however, there are certain factors that contribute to low cost. For
example, in a one-story house if there is a common wall between the
bathroom and kitchen, with the sink mounted on the kitchen side and
the bathroom fixtures grouped on the other side, a minimum amount of
material and labor is required. In a two-story house if the bathroom
can be immediately above the kitchen, the same is true.

Coupled with low cost are the added advantages in this arrangement of
reduced pipe freezing hazards and short hot water lines, so that only
a small amount of water need be drawn before hot water is available
at the sink or lavatory.

When all the bathroom fixtures are grouped on one wall, regardless of
where the bathroom is located, a minimum amount of piping is required
to connect the fixtures. Note figure 99b.

Other factors affecting installation cost are (1) distance from the
water supply line to the bathroom, and (2) distance to the septic tank.

Full advantage can seldom be taken of all of the cost saving features,
since a balance must be reached between low cost and convenience of the
bathroom as it relates to the rest of the house plan, as well as relation
of the sink to a well planned kitchen. However, they are worth keeping
in mind.

Selection of Bathroom Space

If a new house is to be constructed there are many good house plans available with the bathroom properly located and the element of cost considered.

Selection of space in a house already constructed, particularly the older ones, often constitutes a genuine problem. Large rooms, poor arrangement of rooms, high ceilings, lack of furnace heat, a large family, and presence of elderly people or a semi-invalid in the home are all factors for consideration.

As little as 25 square feet of floor space can be used for a bathroom. That amount of space can often be found in a large closet or end of a hall. A larger space is usually desirable if it can be made available. Many homes with large rooms can utilize a portion of one or possibly two rooms, as shown in figure 97.

If a full sized room is to be converted into a bathroom it may be desirable to reduce the size, particularly if the room is large. This is often accomplished by partitioning off the more desirable portion of the room for the bathroom and converting the remainder into useful space, such as a clothes closet or storage room.

If a bathroom is also used as a dressing room, it should be fairly good sized to provide for a dressing table, chair, etc.

Figure 97. A Bathroom from Parts of Two Rooms

Farm families in homes with no central heating plant often confine their living in winter to two or three rooms, which may include the kitchen, living room, dining room, or bedroom. Since heating is also confined to that portion of the house, the bathroom must of necessity be directly connected with one of the actively used rooms. The arrangement is not good but probably the best that can be had under existing circumstances. If a room must be added for the purpose, a complete foundation should be laid to prevent air from circulating under the floor.

In selecting the bathroom space the following conditions should be met as nearly as possible:

1. Locate near bedrooms.
2. Have one entrance, preferably from a hall.
3. Have minimum outside wall exposure.
4. Install pipes on inside walls.

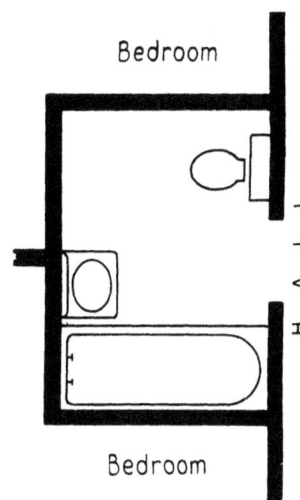

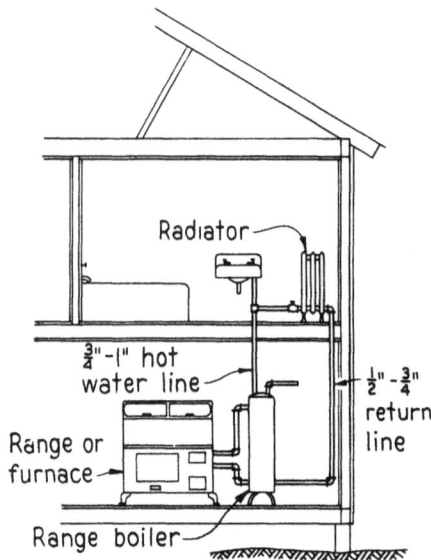

Figure 98. Radiator Installed with Range Boiler for Supplying Supplemental Bathroom Heat

5. Provide for access to pipes in wall.
6. Provide for a linen closet.
7. If possible have at least one window.
8. Keep room reasonably small so it can be easily heated.
9. Provide some means of supplemental heat if there is no central heating plant. If electric rates permit, an electric heater is desirable. If a range boiler is used in connection with a furnace or kitchen range, and the range boiler is in a position lower than the bathroom the arrangement in figure 98 is practical.

Bathroom Arrangement

Basic Plans

There are six basic bathroom plans for fixture arrangement, as shown in figure 99. The arrangement is adaptable to rooms of any size or shape.

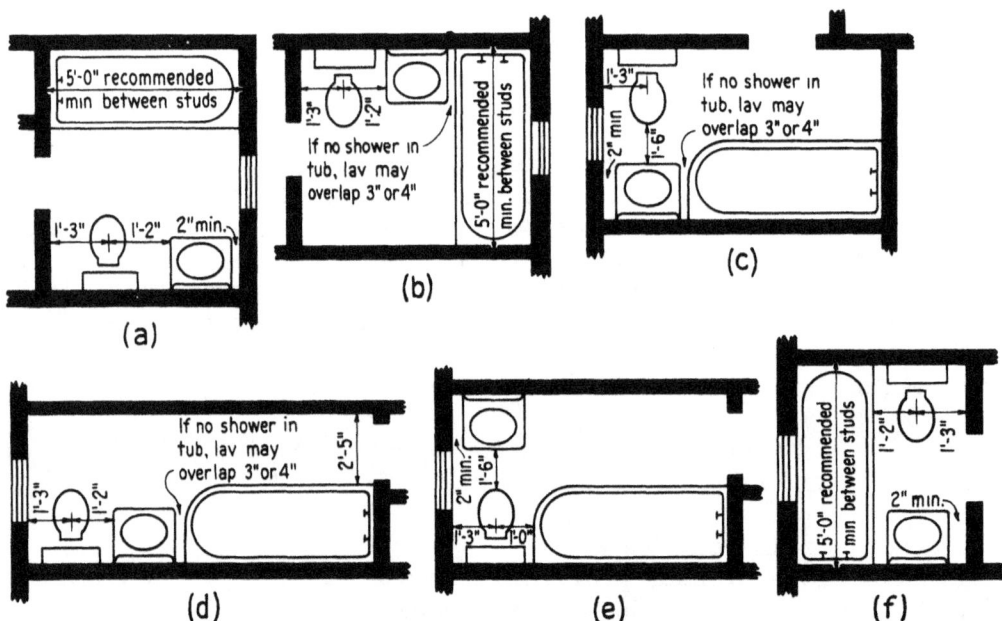

Figure 99. Six Basic Bathroom Plans with Minimum Dimensions

The dimensions given are considered minimum. If more space is available
it is well to place the fixtures farther apart.

The size and shape of the room affect the bathroom plan. Likewise, position of the stack and entrance of supply piping are determing factors.
The closet should sit as close as possible to the stack.

In old houses it is sometimes difficult to conceal the plumbing. If a
new wall is to be constructed, 6-inch studding should be used to provide
for a 3-inch stack. A linen or clothes closet can sometimes be used, or
the stack and service pipes may be mounted in one corner and furred in
(figure 100).

For pleasing appearance and convenience
the following suggestions may be of value.

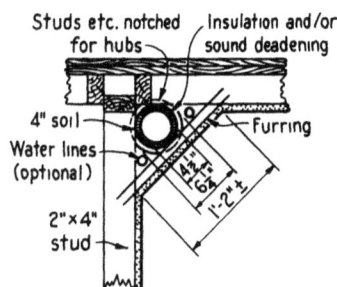

Figure 100. Pipes Concealed
by Furring In

Window

1. The window is best located at
    right angle with lavatory.
2. It is least desirable over tub:
    (a) It is difficult to open
        window.
    (b) Open window allows air to
        strike bather.
    (c) It is not well adapted to
        shower installation.
3. If window must be over a fixture, the closet is first choice,
    with window 4-1/2 to 5 feet from floor.

Door

1. There should be only one bathroom entrance.
2. The door may be as small as 2 feet wide.
3. If space is limited the door may swing "out" instead
    of "in."
4. It should be located so as not to strike a person
    using a fixture.
5. It is desirable for the door, when open, to conceal
    the closet fixture.

If fixtures are on one wall the lavatory should be in the center to provide elbow room.

Ample light should be provided with a fixture over the mirror or one on
each side.

A convenience outlet should be provided beyond reach of anyone standing
in the tub.

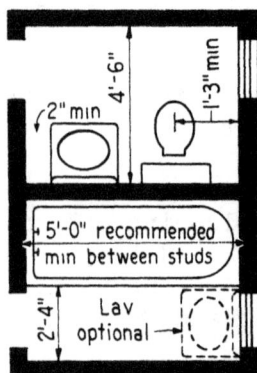

Figure 101. A Two Compartment Bathroom

Moldings and trim should be eliminated or reduced to a minimum.

For large families a two-compartment bath is particularly convenient and relatively inexpensive (figure 101).

Bathroom Accessories

Bathroom accessories commonly consist of towel bars, tumbler and toothbrush holder, soap dishes, toilet paper holder, mirror over lavatory, and sometimes a glass shelf over the lavatory.

A built-in medicine cabinet with mirrored door over the lavatory is very desirable.

Among the desirable things not found in most farm bathrooms are grab bars and full length mirrors.

Wall Coverings

With colored fixtures available on the market there is a marked tendency to add color also in wall and floor coverings. Materials such as marble and linoleum provide color in themselves and should be carefully selected for proper color blending. Keene cement plaster and composition boards are among the materials that are painted so that color combinations can be changed if desired. The more common wall coverings are:

1. Waterproof wallpaper--satisfactory, particularly if used on the upper wall with a hard surface wainscoting for the lower wall.
2. Keene cement--used as a plaster and commonly marked to appear like tile. It can be painted, is water resistant, and gives good service.
3. Composition boards--available in large sizes, requiring few joints. Nonporous types are water resistant, easily erected and can be painted. Joints are usually covered with metal strips.
4. Linoleum--very satisfactory, but special care is required to prevent moisture from getting behind it. When used for wainscoting a special paste is applied to the wall, 3/4-inch brads

Single light
(centered)

Mirror and
medicine cab.

If side lights, use pair preferably with large area diffusers.

1'-6" min.

7" min.

Convenience
outlet

Recessed
shelf

6'-0" min.

Soap

Tooth brush
holder

Towel bars

2'-7" normal

Curtain rod

Shower head

6"

Permit no projection
beyond tank or flush
valve less than 4'-6"
from floor

Robe hook

Adults 5'-9", Women only 5'-6", Children only 5'-0"

Shower control

Towel bar
(omit with shower)

Min 6'-0"- usually 6'-6"

Reach 30" average

5'-0" to 5'-6"

Approximately

1'-0"   1'-0"

4'-0" min.

Vertical
grab bars

Soap

Toilet paper
holder

2'-0" min.
3'-0" max.

4'-6"

2'-4" to 2'-6"

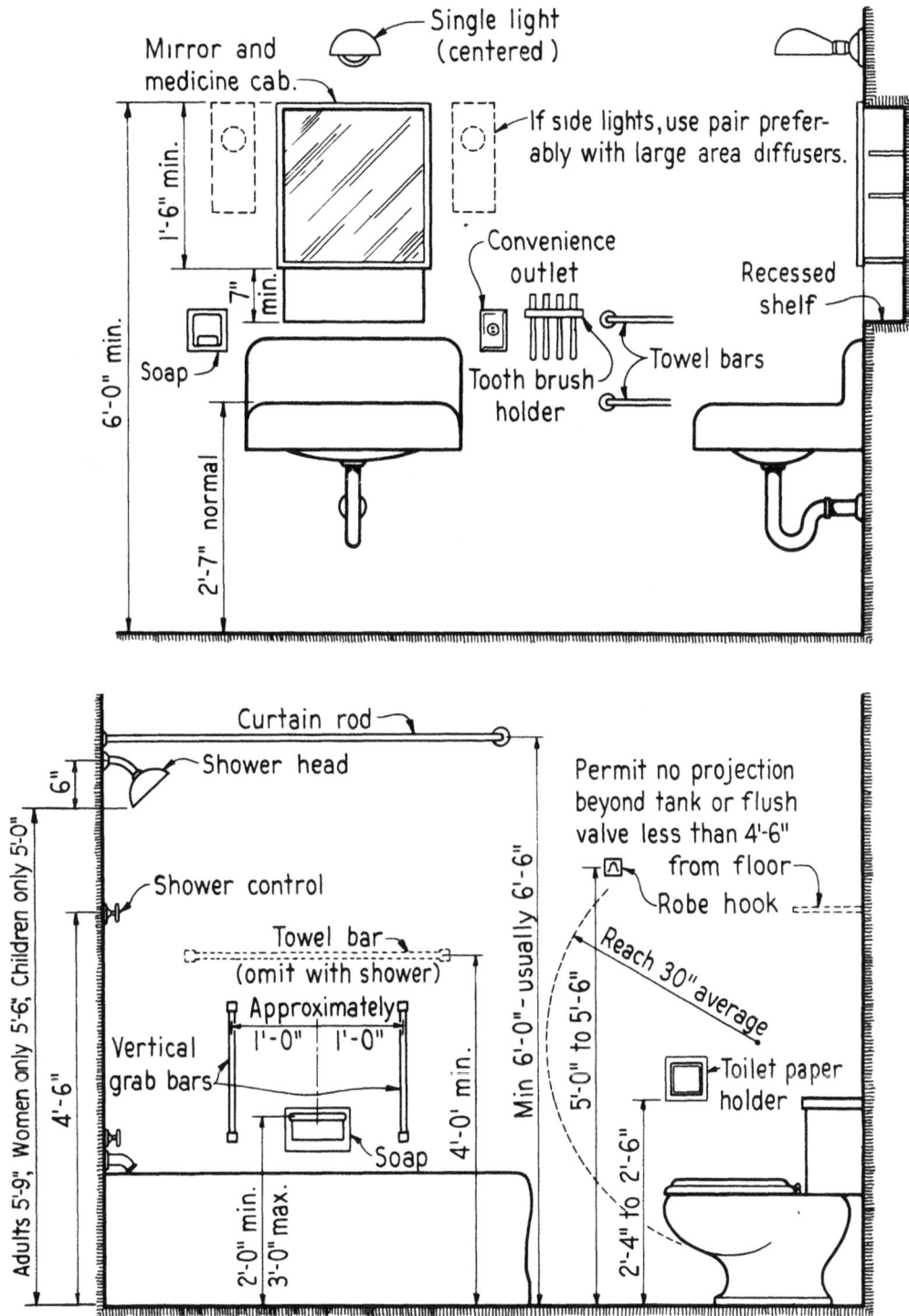

Figure 102.   Suggested Placement and Measurements
for Bathroom Accessories

are nailed along the top, and a special strip is used on the floor edge.

5. Marble--expensive but very satisfactory. It is colorful, sanitary, and requires little upkeep.
6. Tile--satisfactory and colorful. There is a tendency for sections of it to come loose unless carefully laid.

Among the less common materials are structural glass, steel, bakelite, cork, and asbestos board.

Floor Coverings

Wood floors without covering are common but difficult to clean and keep neat appearing in the presence of moisture and soap. Most paints and varnishes are not sufficiently waterproof to protect the floor fully.

1. Linoleum--very satisfactory if laid with waterproof cement to keep moisture from getting under it.
2. Tile--ceramic tile is satisfactory and can be obtained in colors. Quarry tile is also satisfactory but not generally used. Glazed tile will not wear well.
3. Marble--satisfactory as long as the nonporous variety is used. Porous types will stain and change color.

Other less common coverings are rubber, cork, asphalt, tile, and cement

REFERENCES

The following publications are valuable for further study of farm pumps, plumbing, and sewage disposal.

Books

Rural Water Supply and Sanitation, by Forrest B. Wright. Published by John Wiley and Sons Inc. New York, 1939.

Bulletins and Leaflets

Farm Water Works and Sewage Systems, by T. B. Chambers and M. L. Nichols. Ala. Poly. Inst. Ext. Circ. 80, reprint (i.e., rev.), 39 pp. illus. Auburn, 1940.

Water Systems for the Farm Home, by E. L. Arnold. Ark. Agr. Col. Ext. Circ. 416, 25 pp. illus. 524 Post Office Bldg., Little Rock, 1938.

Farm Home Water Systems, Source and Location of Water Supplies, Inexpensive Water Systems, Cisterns and Septic Tanks, by Exine Davenport. Colo. State Col. Ext. Bul. 352-A, 16 pp illus. Fort Collins, 1938.

Water and Sewerage Systems for Florida Rural Homes, by Frazier Rogers. Fla. Univ. Agr. Ext. Serv. Bul. 91, 20 pp diagrs. Gainesville, 1937.

Water and Plumbing Systems for Farm Homes, by E. W. Lehmann and F. P. Hanson. Ill. University Agr. Expt. Sta. and Ext. Serv. Circ. 303, reprint, 20 pp illus. Urbana, 1936.

Discussion Outline on Community Action for Safe and Convenient Water in the Home, by D. E. Lindstrom and R. R. Parks. Ill. Univ. Agr. Ext. Serv. RSE-51, 5 pp. Urbana, 1937. Mimeographed.

Farm and Home Water Systems, by C. H. Van Vlack. Iowa State Col. Ext. Circ. 256, 12 pp. illus. Ames, 1939.

Electrically Operated Water Systems for Farms, by J. B. Brooks. Ky. Agr. Col. Ext. Circ. 319, 31 pp. illus. Lexington, 1938.

Running Water for Farm and Home, by M. G. Huber. Maine Agr. Col. Ext. Bul. 250, 31 pp. illus. Orono, 1938.

The Hydraulic Ram, by W. H. Sheldon. Mich. State Col. Ext. Bul. 171, 9 pp. diagrs. East Lansing, 1936.

Farm Wells, by A. G. Tyler. Minn. Univ. Agr. Ext. Div., Agr. Engin. News Letter No. 51, 1 p. University Farm, St. Paul, 1936

Effect of Drainage on Water Levels of Farm Wells, by D. G. Miller. Minn. Univ. Agr. Ext., Agr. Engin. News Letter No. 87, 1 p. illus. University Farm, St. Paul, 1939.

An Electric Water System and Plumbing for the Farm, by F. M. Hunter. Miss. State Col. Ext. Circ. 94, 10 pp. illus. State College, 1937.

Water and Sewage Disposal for Farm Homes, by J. C. Wooley and others. Mo. Agr. Col. Ext. Circ. 401, 15 pp. diagrs. Columbia, 1939.

Farm Water Systems, by J. C. Wooley and others. Mo. Agr. Col. Ext. Circ. 413, 27 pp. illus. Columbia, 1940.

The Domestic Water Supply on the Farm, H. E. Murdock. Mont. State Col. Ext. Bul. 172, 24 pp. illus. Bozeman, 1939.

A Shallow-well Water System, by C. N. Turner. N. Y. Agr. Col. (Cornell) Ext. Bul. 392, 15 pp. illus. Ithaca, 1938.

A New Deep-well Water System, by C. N. Turner. N. Y. Agr. Col. (Cornell) Ext. Agr. Engin. Seasonal Suggestions (unnumb.) 1 p. diagr. Ithaca, 1939.

A Complete Hand-power Water System for the Farm Home, by D. S. Weaver. N. C. State Col. of Agr. Ext. Folder 38, 5 pp. illus. State College Station, Raleigh, 1937.

Simple Water Systems for the Farm Home, by D. S. Weaver. N. C. State Col. of Agr. Ext. Folder 37, 8 pp. diagrs. State College Station, Raleigh, 1937.

Farm Home Water Systems at Low Cost, by L. E. Holman and H. F. McColly. N. Dak. Agr. Col. Ext. A. E. 31, 11 pp. illus. State College Station, Fargo, 1938. Mimeographed.

Water Systems for the Farm Home, by J. W. Carpenter, Jr., and C. V. Phagan. Okla. Agr. Col. Ext. Circ. 245. 24 pp. illus. Stillwater, 1933.

Farm Water Systems, by E. H. Davis. Oreg. State Col. Ext. Circ. 337, 12 pp. illus. Corvallis, 1939. Mimeographed.

Improved Water Systems for Tennessee Farm Homes. Tenn. Agr. Col. Ext. Spec. Circ. 53, 8 pp. Knoxville, 1937. Mimeographed.

Panel discussion: Running Water in the Farm Home, by Lillian
L. Keller and others. Tenn. Agr. Col. Ext. Spec. Circ. 55,
7 pp. Knoxville, 1937. Mimeographed.

A Water System for the Farm Home, by G. W. Boyd. Wyo. Agr.
Col. Ext. Circ. 69, 15 pp. illus. Laramie, 1939.

Sewerage Systems for Farm Homes and Unsewered Communities,
by Felix J. Underwood and H. A. Kroeze. Mississippi State
Board of Health, Health Bulletin No. 19. Jackson.

Sewage Disposal for Rural Dwellings, issued by the Dept. of
Health, Division of Sanitary Engin., State of Ohio, and The
Agricultural Ext. Serv. Dept. of Agr. Engineering, College
of Agr., the Ohio State Univ. No. 112, reprint. Columbus,
June 1934.

Septic Tanks. Virginia Health Bulletin, Vol. XXVIII, July
1936 Supplement No. 7, Richmond, Va.

Disposal of Farm Sewage, by G. O. Hill. Coop. Ext. Work in
Agr. and Home Ec., Purdue Univ. Ind. Ext. Bul. No. 165.
(reprint, April 1936)

Farm Sewage Disposal, by Paul R. Hoff and H. J. Young.
Univ. of Neb. Agr. Col. Ext. Serv. and U. S. Dept. of Agr.
Cooperating. Ext. Circ. No. 703. Lincoln, Feb. 1931.

Sanitation Applied to the Rural Home, by Dr. E. L. Bishop,
Dir., Bureau of Rural Sanitation, State Board of Health,
Nashville, Tenn. Publication 81, May 1920.

The Septic Tank System for Home Sewage Disposal, by Earle G.
Brown, M. D. Bulletin of Kansas State Board of Health, in
coop. with School of Engin. and Arch., Univ. of Kan., and
Ext. Serv. Vol. 14, No. 9, August 1936. Engin. Bul. No. 18,
Univ. of Kansas, Lawrence.

A Concrete Septic Tank for the Farm, by Everett H. Davis.
Ext. Circ. 333, Fed. Coop. Ext. Serv. Oregon State College,
Corvallis, July 1939.

Septic Tanks for Sewage Disposal, by Earl. G. Welch and James
B. Kelley. Univ. of Ky., Col. of Agr. Ext. Circ. No. 131
(Revised) Lexington, Ky. June 1931.

Sewage Disposal for Farm Homes, by S. A. Witzel and F. R.
King. Ext. Serv., College of Agr., Univ. of Wisconsin,
Madison. Stencil Circ. 205. Nov. 1938.

Water Supplies for Suburban and Country Homes, Dug Well, Supplies. Va. State Dept. of Health, Richmond, Va.

Good Water for Farm Homes, by A. W. Freeman. Public Health Service. Public Health Bulletin No. 70, May 1915. U. S. Govt. Printing Office, Washington, 1916.

Springs. Va. State Dept. of Health, Richmond, Va.

Ground-water Supplies. Public Health Service. Supplement No. 124 to the Public Health Reports. U. S. Govt. Printing Office, Washington, 1937.

Rural Water Supplies, Ohio Dept. of Health, Div. of Sanitary Engineering, Columbus, Ohio, 1937.

Soft Water for the Home, by A. M. Buswell and E. W. Lehmann. Univ. of Ill., Agr. Expt. Sta. Circ. 393. Urbana, 1932.

Wisconsin State Well Drilling Sanitary Code, Wisconsin State Board of Health, 1936, First Issue. 30¢ per copy to non-residents of Wisconsin. Madison.

Water Softening for the Home, by Lindon J. Murphy. Eng. Ext. Serv. of Iowa State College, Bulletin No. 105, Ames, Iowa. October 1, 1930.

Staple Vitreous China Plumbing Fixtures. Bureau of Standards. Commercial Standard CS20-30, issued June 26, 1930. U. S. Govt. Printing Office, Washington, 1930. Price 10¢.

Water Supply and Plumbing for the Farm Home, by Hobart Beresford. Ext. Bulletin No. 95 of Univ. of Idaho, College of Agr., Moscow, December 1934.

Colors for Sanitary Ware. Bureau of Standards. Commercial Standard CS30-31. Issued Jan. 20, 1932. U. S. Govt. Printing Office, Washington. Price 20¢.

Simple Plumbing Repairs in the Home, by George M. Warren. U. S. Dept. of Agriculture, Farmers' Bulletin No. 1460. Revised October 1936.

Sewage Disposal for the Home. Ga. State Board of Health, Engineering Bulletin No. 3, Div. of San. Engineering, Atlanta, Ga., September 1928.

Private Water Supplies and Private Sewage Disposal Systems in Wisconsin. Wis. State Board of Health, Bureau of Plumbing and Domestic San. Engineering, 1937. Reprinted from the 1937 issue of the Wis. State Plumbing Code. Madison.

Water Supplies for Rural Homes. Miss.State Board of Health, Bureau of San. Engineering. Health Bulletin No. 16. Jackson.

Ground-water Supplies. Public Health Service. Supplement No. 124 to the Public Health Reports. U. S. Govt. Printing Office, Washington, 1937. Price 5¢.

Sewage Disposal for the Farm Home, by R. C. Kelleher and E. W. Lehmann. College of Agr. and Agr. Experiment Station of Univ. of Illinois, Circular 336. Urbana, March 1929.

Instructions for Installing Modern Plumbing Systems. Sears, Roebuck and Co.

APPENDIX

USEFUL RULES AND TABLES

RULES

To convert inches vacuum into feet suction, multiply by 1.13.

To reduce pounds pressure to feet head, multiply by 2.3.

To reduce heads in feet to pressure in pounds, multiply by .43.

Friction of liquids in pipes increases as the square of the velocity.

Doubling the diameter of a pipe increases its capacity four times.

To find the area of a pipe, square the diameter and multiply by .7854.

A cubic foot of water weighs 62-1/2 pounds and contains 1728 cubic inches or 7-1/2 U. S. gallons.

The gallons per minute which a pipe will deliver equals .0408 times the square of the diameter, multiplied by the velocity in feet per minute.

To find the capacity of a pipe or cylinder in gallons, multiply the square of the diameter in inches by the length in inches and by .0034.

To find the capacity of a pipe or cylinder in cubic inches, multiply the square of the diameter in inches by the length in inches and by .7854.

To find the discharge from any pipe in cubic feet per minute, square the diameter and multiply by the velocity in feet per minute and by .00545.

U. S. gallon of water weighs 8-1/3 pounds and contains 231 cubic inches.

To find the capacity of a given tank or cistern in U. S. gallons, square the diameter (in feet), and multiply by .7854, multiply by the height in feet, and by 7.48.

TABLE VIII

FRICTION OF WATER IN PIPES

Loss of Head in **FEET** Due to Friction, per 100 Feet of 15 year old **Ordinary Iron Pipe**

U.S. Gal. per min.	½″ Pipe Vel.	Fric.	¾″ Pipe Vel.	Fric.	1″ Pipe Vel.	Fric.	1¼″ Pipe Vel.	Fric.	1½″ Pipe Vel.	Fric.	2′ Pipe Vel.	Fric.	2½″ Pipe Vel.	Fric.	3″ Pipe Vel	Fric.
1	1.05	2 1														
2	2.10	7.4	1 20	1 9												
3	3 16	15 8	1 80	4 1	1 12	1 26										
4	4 21	27 0	2 41	7 0	1 49	2 14	0 86	0 57	0 63	0 26						
5	5 26	41 0	3 01	10 5	1 86	3 25	1 07	0 84	0 79	0.39						
0	10 52	147 0	6 02	38 0	3 72	11 70	2 14	3.05	1 57	1 43	1 02	0 50	0.65	0 17	0 45	0 07
5			9.02	80 0	5 60	25 00	3 20	6 50	2 36	3 00	1 53	1 00	0 98	0 36	0 68	0.15
0			12 03	136 0	7 44	42 00	4 29	11 10	3 15	5 20	2 04	1 82	1 31	0 61	0 91	0 25
5					9 30	64 00	5 36	16 60	3 94	7 80	2 55	2 73	1.63	0 92	1 13	0 38
0					11 15	89 00	6 43	23 50	4 72	11 00	3 06	3 84	1 96	1 29	1 36	0 54
5					13 02	119.00	7 51	31 20	5 51	14 70	3.57	5 10	2 29	1 72	1 59	0.71
0					14 88	152 00	8 58	40 00	6 30	18 80	4 08	6 60	2 61	2 20	1 82	0 91
5							9 65	50 00	7 08	23 20	4 60	8 20	2 94	2 80	2 05	1 15
0							10 72	60 00	7 87	28 40	5 11	9 90	3.27	3 32	2 27	1 38
0									11 02	53 00	7 15	18 40	4 58	6 20	3 18	2 57
0									14 17	84 00	9 19	29 40	5 88	9 80	4 09	4 08
0											10 21	35 80	6 54	12 00	4 54	4 96
0											12 25	50 00	7 84	16 80	5 45	7 00
0											14 30	67 00	9 15	22 30	6 35	9 20
0											16 34	86 00	10 46	29 00	7 26	11 80
0													11 76	35 70	8 17	14 80
0													13 07	43 10	9 08	17 80
0													14 38	52 00	9 99	21 30
0													15 69	61.00	10 89	25 10
0													16 99	70.00	11 80	29 10
0													18 30	81.00	12 71	33 40
0													19.61	92 00	13 62	38 00

U.S. Gal per min.	4″ Pipe Vel.	Fric.	5″ Pipe Vel.	Fric.	6″ Pipe Vel.	Fric.	8″ Pipe Vel.	Fric.	10″ Pipe Vel	Fric.	12″ Pipe Vel.	Fric.	14″ Pipe Vel.	Fric.	15″ Pipe Vel.	Fric.
0	1 02	0 22														
5	1 17	0 28														
0	1 28	0 34														
0	1 79	0 63	1 14	0 21												
5	1 92	0 73	1 22	0 24												
0	2 55	1 23	1 63	0 39	1 14	0 14										
0	3 06	1 71	1 96	0 57	1 42	0.25										
5	3 19	1 86	2 04	0 64	1 48	0 28										
0	3 84	2 55	2 45	0 88	1 71	0 32										
5	4 45	3 36	2 86	1 18	2 00	0 48										
0	5 11	4 37	3 27	1 48	2 28	0 62										
5	6 32	6 61	3 67	1 86	2 57	0 74										
0	6 40	6 72	4 08	2 24	2 80	0 92	1 60	0 22								
5	7 03	7 99	4 50	2 72	3 06	1 15	1 73	0 27								
0	7 66	9 38	4 90	3 15	3 40	1 29	1 90	0 36								
0	8 90	12 32	5 72	4 19	3 98	1 69	2 20	0 41								
0	10 20	15 82	6 54	5 33	4 54	2 21	2 60	0 56								
0	11 50	19 74	7 35	6 65	5 12	2 74	2 92	0 64	1 80	0 21						
0	12 30	22 96	7 88	7 22	5 55	3 21	3 10	0 79	1 94	0 25						
0	12 77	24 08	8 17	8 12	5 60	3 26	3 20	0 81	2 04	0 28	1 42	0.11				
0			8 99	9 66	6 16	3 93	3 52	0 98	2 25	0 33	1 57	0 14				
0			9 80	11 34	6 72	4 70	3 84	1 16	2 16	0 39	1 71	0 15				
0			10 62	13 16	7 28	5 50	4 16	1 34	2 66	0 46	1 85	0 19	1 37	0 09		
0			11 11	15 12	7 84	6 38	4 46	1 54	2 86	0 52	2 00	0 22	1 47	0 10		
0			12 26	17 22	8 50	7 00	4 80	1 74	3 06	0 59	2 13	0 24	1 58	0 11		
0					9 08	7 90	5 12	1 97	3 28	0 67	2 27	0 27	1 68	0 13		
0					9 58	8 75	5 48	2 28	3 48	0 75	2 41	0 31	1 79	0 14		
0					10 30	10 11	5 75	2 46	3 68	0 83	2 56	0 34	1 89	0 16		
0					10 72	10 71	6 06	2 87	3 88	0 91	2 70	0 35	2 00	0 16	1 73	0 12
0					11 32	12 04	6 40	3 02	4 08	1 01	2 84	0 41	2 10	0 19	1 82	0 14
0					12 50	14 31	7 03	3 51	4 50	1 20	3 13	0 49	2 31	0 23	2 00	0 16
0					13 52	16 69	7 67	4 26	4 91	1 46	3 41	0 57	2 52	0 26	2 18	0 19
0							9 60	6.27	6 10	2 09	4.20	0 85	3 15	0 39	2 73	0 28
0							12 70	10 71	8 10	3 50	5 60	1 43	4 20	0 66	3 64	0 17

In figuring Vertical Suction Depth on Shallow Well Pumps with long horizontal Suction Pipe, the loss by friction in water passing through this long Pipe must be deducted from the maximum suction limit of the Pump as under ordinary conditions.

Vel.—Velocity feet per second.   Fric.—Friction head in feet

Friction of Water in 90° Elbows or Tees

Size of Elbow, Inches …	½	¾	1	1¼ to 2	2½	3	4	5	6	8	10	12	14	15
Equivalent Number of Feet Straight Pipe	5	6	6	8	11	15	16	18	18	24	30	40	54	55

For Globe Valves add 50% to above.

TABLE IX

PRESSURE OF WATER PER SQUARE INCH AND FEET HEAD

	Feet Head of Water and Equivalent Pressure						Lbs., Pressure and Equivalent Ft. Head of Wa				
Feet Head	Lbs. per Sq. Inch	Feet Head	Lbs. per Sq. Inch	Feet Head	Lbs. per Sq. Inch	Lbs. per Sq. Inch	Feet Head	Lbs. per Sq. Inch	Feet Head	Lbs. per Sq. Inch
1	43	60	25 99	200	86 62	1	2 31	40	92 36	170
5	2 17	100	43 31	300	129 93	5	11 54	80	184 72	225
10	4 33	150	64 96	600	259 85	10	23 09	125	288 62	350
20	8 66	160	69 29	700	303 16	15	34 63	130	300 16	375
30	12 99	170	73 63	800	346 47	20	46 18	140	323 25	400
40	17 32	180	77 96	900	389 78	25	57 72	150	346 34	500
50	21 65	190	82 29	1000	433 09	30	69 27	160	369 43	1000

TABLE X

FLOW OF WATER PER MINUTE, DIFFERENT FIXTURES

Bath ....10 Gallons p
Lavatory .... 5 Gallons p
Tank Closets .... 5 Gallons p
Valve Closets .... 30 Gallons p
Shower .... 5 Gallons p
Sink ....10 Gallons p
Laundry Tub ....10 Gallons p
Garden Hose (Sprinkling Nozzle ¾") .... 5 Gallons p
Continuous Drinking Fountain ....1½ Gallons p

TABLE XI

CAPACITIES OF WATER PIPING IN BUILDING—LENGTH 100 FT.
GALS. PER MIN.

Size Pipe	½	¾	1	1¼	1½	2	2½	3
Pressure								
17 lbs	3 2	9 1	18.7	33 5	51.6	106	200	290
30 lbs.	5	14	28	52	78	160	308	436
40 lbs	6	16	33	60	90	184	350	504
50 lbs.	6 5	17 5	37	70	101	206	390	564
60 lbs	7	19 5	40	76	110	226	430	617
75 lbs	7.5	22	45	85	123	253	480	690
100 lbs	9	25	52	99	142	292	558	797

TABLE XII

Weight of Water Contained in One Foot Length of Pipe of Different $

Size	Pounds	Size	Pounds	Size	Pou
½	086	2	1 372	4	5
1	343	2½	2 159	4½	6
1¼	537	3	3 087	5	8
1½	774	3½	4 211	6	12

Multiply by number of feet high or head

TABLE XIII

Number Gallons in Cistern and Tanks

Depth in Feet	DIAMETER IN FEET											
	5	6	7	8	9	10	11	12	13	14	15	16
5	725	1,060	1,440	1,875	2,380	2,925	3,550	4,237	4,960	5,765	6,698	7,520
6	870	1,270	1,728	2,250	2,855	3,510	4,260	5,084	5,952	6,918	8,038	9,024
7	1,015	1,180	2,016	2,625	3,330	4,112	4,970	5,931	6,944	8,071	9,378	10,528
8	1,160	1,690	2,304	3,000	3,805	4,680	5,680	6,778	7,936	9,224	10,718	12,032
9	1,305	1,900	2,592	3,375	4,280	5,265	6,380	7,625	8,928	10,377	12,058	13,536
10	1,450	2,110	2,880	3,750	4,755	5,850	7,100	8,472	9,920	11,530	13,398	15,040

TABLE XIV

Capacity of Steel Tanks

Diam. Inches	Gals. per ft. Length	Diam. Inches	Gals. per ft. Length	Diam. Inches	Gals. per ft. Length	Diam. Inches	Gals. per ft. Length	Diam. Inches	Gals. per ft. Length	Diam. Inches	Gals. per ft. Length	Diam. Inches	Gals. per ft. Length
12	5 87	17	11 79	22	19 75	27	29 74	32	41 78	37	55 86
13	6 89	18	13 22	23	21 58	28	31 99	33	44 43	38	58 92
14	8 00	19	14 73	24	23 50	29	34 31	34	47 16	39	62 06
15	9 18	20	16 32	25	25 50	30	36 72	35	49 98	40	65 28
16	10 44	21	17 99	26	27 58	31	39 21	36	52 88		

TABLE XV

COMMON PRESSURE TANK SIZES

Capacity	Size	Approx. Weight		Capacity	Size	Approx. Weight
				HORIZONTAL		
12 Gallon Vertical	12 x 24"	37 lbs.		525 Gallon Horiz.	36 x 120"	788 lbs.
42 Gallon Vertical	16 x 48"	95 lbs.		720 Gallon Horiz.	42 x 120"	1205 lbs.
42 Gallon Vertical	20 x 32"	112 lbs.		1000 Gallon Horiz.	42 x 168"	1595 lbs.
82 Gallon Vertical	20 x 60"	160 lbs.		1500 Gallon Horiz.	48 x 192"	2,05 lbs.
120 Gallon Vertical	24 x 60"	214 lbs.		2000 Gallon Horiz.	48 x 258"	2650 lbs.
220 Gallon Vertical	30 x 72"	444 lbs.		3000 Gallon Horiz.	60 x 238"	5500 lbs.
315 Gallon Vertical	36 x 72"	555 lbs.				

TABLE XVI

MINIMUM WIRE SIZE BETWEEN POWER SOURCE AND MOTOR.

Horsepower and Voltage of Motor	One-Way Distance to Motor							
	100'	150'	200'	250'	300'	400'	500'	1000'
1/4 hp., 115-volt	12	12	12	12	12	10	10	8
1/3 hp., 115-volt	12	12	12	12	10	10	8	6
1/2 hp., 115-volt	12	12	10	10	10	8	8	4
3/4 hp., 115-volt	12	12	10	10	8	8	6	4
1 hp., 115-volt	10	10	8	8	8	6	6	2
1-1/2 hp., 115-volt	10	8	8	8	6	6	4	2
2 hp., 115-volt	10	8	8	6	6	4	2	0
1/4 hp., 230-volt	12	12	12	12	12	12	12	12
1/3 hp., 230-volt	12	12	12	12	12	12	12	12
1/2 hp., 230-volt	12	12	12	12	12	12	12	10
3/4 hp., 230-volt	12	12	12	12	12	12	12	10
1 hp., 230-volt	14	14	14	14	12	12	10	8
1-1/2 hp., 230-volt	14	14	14	12	12	10	10	8
2 hp., 230-volt	14	14	12	12	10	10	8	6
3 hp., 230-volt	12	12	10	10	10	8	8	4
5 hp., 230-volt	8	8	8	8	6	6	6	2
7-1/2 hp., 230-volt	6	6	6	6	6	4	4	2
10 hp., 230-volt	4	4	4	4	4	4	2	0

USE OF A WEIR

Where the flow of water is large and therefore not practical for measuring by means of a bucket or tub, make use of a notch in a board, known as a "Weir."

Measure the width of the notch "W" and the height of the water in the notch "H." The height should be measured on a level 2 feet up stream from the notch.

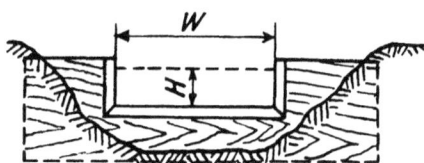

Figure 103.

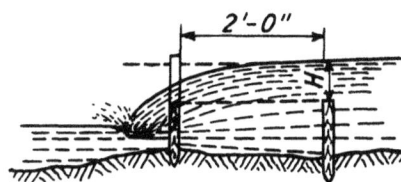

Figure 104.

TABLE XVII

WEIR TABLE FOR DETERMINING THE FLOW OF STREAM

INCHES	0	⅛	¼	⅜	½	⅝	¾	⅞
0.........	0.00	0.01	0.05	0.09	0.14	0.19	0.26	0.32
1.........	0.40	0.47	0.55	0.64	0.73	0.82	0.92	1.02
2.........	1.13	1.23	1.35	1.46	1.58	1.70	1.82	1.95
3.........	2.07	2.21	2.34	2.48	2.61	2.76	2.90	3.05
4.........	3.20	3.35	3.50	3.66	3.81	3.97	4.14	4.30
5.........	4 47	4.64	4.81	4.98	5.15	5.33	5.51	5.69
6.........	5.87	6.06	6.25	6.44	6.62	6.82	7.01	7.21
7.........	7.40	7.60	7.80	8.01	8.21	8.42	8.63	8.83
8.........	9 05	9.26	9.47	9.69	9.91	10.13	10.35	10.57
9.........	10 80	11.03	11.25	11.48	11.71	11.94	12.17	12.41
10.........	12.64	12 88	13.12	13.36	13.60	13.85	14.09	14.34

This table gives the number of cubic feet of water that will pass over a weir 1 inch wide and from 1/8 to 10-7/8 inches in depth. The figures in the first upright column represent whole inches and those in the top horizontal line represent fractional parts of an inch of depth over the weir. The figures in the second upright column indicate the number of cubic feet of water that will flow per minute over the weir for whole inches in depth, and in the succeeding columns, whole inches and the fractions under which they occur. Then the number of cubic feet thus found multiplied by the width of the weir in inches will give the capacity of a stream.

Example: To find the required number of cubic feet of water that will flow over a weir 4-3/4 inches in depth and 30 inches in width, follow down the left-hand column of figures in the table to 4, then across until directly under the 3/4 in the top line, where will be found 4.14; this, multiplied by 30 (width of notch in weir) will give 124, the number of cubic feet of water that passes over the whole weir per minute. To reduce to gallons per minute multiply by 7-1/2, which equals 930 gallons per minute.

HOW TO FIGURE APPROXIMATE AMOUNT
OF WATER A RAM WILL DELIVER

A simple formula for figuring the approximate delivery from a hydraulic ram is as follows:

$$\frac{V \times F \times 40}{E} = D$$

V = gallons per minute of supply water.
F = fall in feet.
E = vertical elevation in feet through which water is to be raised.
D - gallons per hour that ram will deliver.

A SIMPLE ELECTRIC WATER SYSTEM

Figure 105 illustrates a simple water system similar to the type designed at Michigan State College and described in Extension Bulletin No. 69, "A Simple Electric Water System."

The arrangement lacks automatic features but has special merit where a farmer may already have money invested in a pump jack and force pump of the type indicated.

Turning the 3-way switch at either the sink or stock tank starts the motor, and water is pumped directly from the well. If the faucets at both the stock tank and sink are closed, the water is delivered to the overhead storage tank. Fresh water is pumped directly to the stock tank if the stock tank valve is opened and the 3-way switch used to start the motor. Water at the sink is delivered from the storage tank unless the motor is started, in which case water is pumped directly from the well.

The system adapts itself to several modifications:

1. The storage tank can be omitted and an open delivery pipe used at the sink.

2. If it is desired to have water from the storage tank supply the stock tank, the check valve may be moved from the position indicated to a point ahead of tee fitting at the pump.

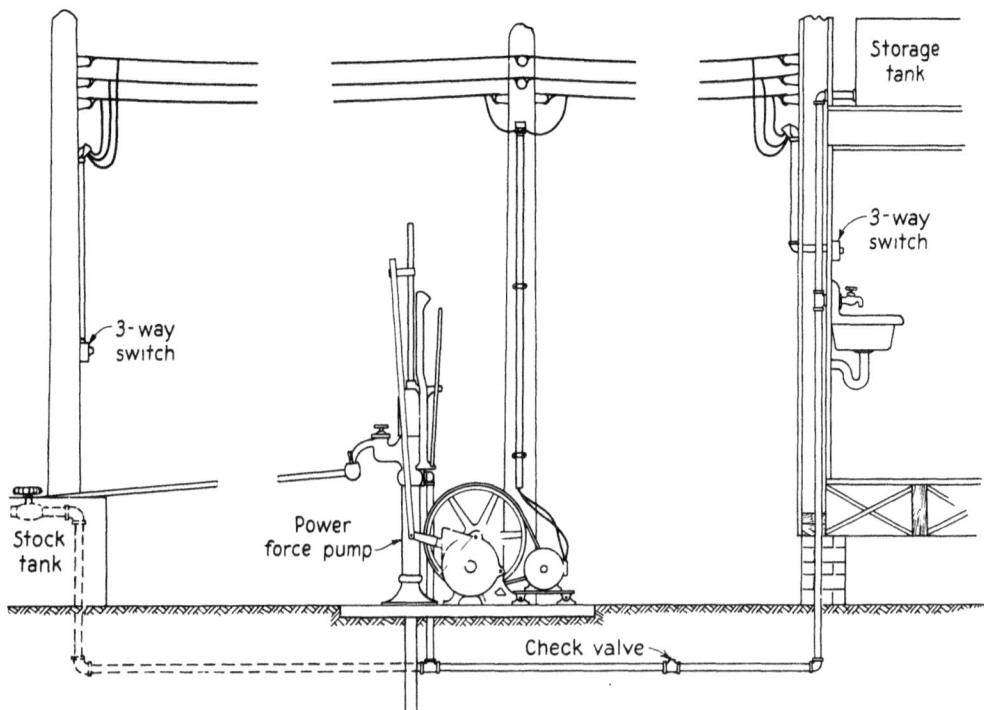

Figure 105. Simple Electric Water System

3. If the pump is a considerable distance from the house and
    stock tank, it may be desirable to install a 4-way switch
    at the pump so that it may be operated from that location
    also.

In most areas provision must be made to keep the pump from freezing.

When a storage tank is used an overflow pipe must be installed. It is
commonly arranged to drain the water down to the sink so as to attract
attention when the tank has filled.

INDEX